Alexander John Wedderbrun

A Popular Treatise of the Extent and Character of Food

Adulterations

Alexander John Wedderbrun

A Popular Treatise of the Extent and Character of Food Adulterations

ISBN/EAN: 9783744646413

Printed in Europe, USA, Canada, Australia, Japan

Cover: Foto ©berggeist007 / pixelio.de

More available books at **www.hansebooks.com**

U. S. DEPARTMENT OF AGRICULTURE.

DIVISION OF CHEMISTRY.

BULLETIN   No. 25.

A POPULAR TREATISE

ON THE

EXTENT AND CHARACTER

OF

FOOD ADULTERATIONS.

BY

ALEX. J. WEDDERBURN,

SPECIAL AGENT.

PUBLISHED BY AUTHORITY OF THE SECRETARY OF AGRICULTURE.

WASHINGTON:
GOVERNMENT PRINTING OFFICE.
1890.

PREFATORY NOTE.

DECEMBER 12, 1889.

SIR: I have the honor to submit herewith the report of Mr. A. J. Wedderburn, special agent of the Department, on the extent and character of food adulterations.

The object of the present bulletin is wholly distinct from that pursued in bulletin No. 13. The investigations, of which the present bulletin is the result, were undertaken for the purpose of collating in popular form well-authenticated facts respecting food adulterations, in order that the people and Congress might have, at least, a general view of the evil which it is hoped Mr. Wedderburn's work may help to remove.

Respectfully,

H. W. WILEY,
Chemist.

Hon. J. M. RUSK,
Secretary.

3

LETTER OF TRANSMITTAL.

ALEXANDRIA, VA., *September* 15, 1889.

Dr. H. W. WILEY,

  Chemist of the U. S. Department of Agriculture:

I have the honor to transmit to you herewith special report upon the Extent of Food Adulterations, a work to which I was duly appointed by commission from the Secretary of Agriculture, March 5, 1889, under act of Congress approved March 2, 1889.

In accordance with my conception of the duty imposed upon me, I spent the greater part of the limited time at my disposal, under terms of my commission, in gathering together important evidence in regard to the extent and character of food adulterations, such information being derived largely from the work of the various State sanitary bodies, official reports, documents, and discussions upon which the laws of the different States were based, the laws of the various States themselves, and the expressions of numerous scientific men on this subject. In preparing a report from the mass of evidence thus collected I have subdivided it into several parts, with some reference to the various sources of information above enumerated. In restricting the report within such limits as would make it available for those for whose perusal it is principally intended, I have necessarily been obliged to exclude a great deal of valuable matter having a direct bearing upon the subject. Enough will be found, however, I am convinced, in the pages of the following report to emphasize in the strongest manner the necessity for such national legislation as was sought during the last session of Congress by Messrs. Conger and Laird of the House Committee on Agriculture, as expressed in their very able reports, as submitted to Congress by order of that committee.

I should add that I have confined myself strictly to the line indicated by the words "report of a popular character," expressed in the commission I hold, and have avoided infringing in any degree upon the scientific work so capably performed by the division under your charge in its investigations into this subject.

I have the honor to be, sir, very respectfully, yours,

ALEX. J. WEDDERBURN,
Special Agent.

CHARACTER OF FOOD ADULTERATIONS.

Adulterations are of two kinds, injurious and non-injurious to health. Adulterations non-injurious to health may be subdivided again into two classes, viz: Those that are simple debasements of well-known dietary articles of which a standard exists, by which to test them, such as flour, grain, lard, wine, sugar, etc. The second subdivision includes such articles as patent medicines, yeast-powders, beer, etc., and presents special difficulties to the investigator from the lack of any such standard and of any fixed rules to govern the formulas. The investigations of the writer show, as will, I think, be conclusively proved by the evidence adduced, that adulteration of our food products is generally and steadily increasing. That most of these adulterations belong to the second class, namely, the non-injurious to health, is a cause for congratulation, and in many cases of adulteration prejudicial to health, the use of such adulterants ought to be charged rather to ignorance than to malice. The result to the consumer, however, is just as grievous, and calls as loudly for his protection from deleterious compounds fraudulently introduced into food, drink, and drugs as though he were the victim of malice.

It may be noted here that among the most poisonous adulterants in use are those used to color and cheapen confectionery and liquors. Now, setting aside for a moment the consideration of this most grievous sort of adulteration, by which the public health is injured, we find what we have styled " non-injurious adulteration " so common that estimates as to the amount of which the American people are annually defrauded in this manner are simply appalling. Several features of this great and growing evil demand special consideration. First, this fraud bears most heavily upon the uneducated and the poor. While the poor man is compelled to patronize cheap stores, and in his struggle for existence and his endeavor to provide the *quantity* necessary to supply the daily wants of himself and family, is driven to purchase cheap goods, the rich man can to a certain extent protect himself by confining his dealings to the most reputable tradesman and paying the highest prices.

Again, we find that adulteration of many of our food products results in cheapening the product of the farm, thus lessening the profits of the husbandman, and robbing both consumer and producer. It must not be forgotten, too, that even though adulterated with mat-

7

ter not positively injurious to health, such food, drugs, or liquors can not be as nutritious and wholesome as the pure articles, and especially important does this feature of adulteration become in the matter of drugs used to prevent or cure disease. To be fed on debased and poisoned food, tainted or diseased meat, until the body sickens, is surely bad enough, without the efforts of the physician to prevent or allay disease being frustrated, by his inability to secure unadulterated drugs and remedies fitted to do his work.

OUR EXPORT TRADE.

Our export trade, of which so large a proportion consists of agricultural products, is also suffering from the same cause, and here again a heavy burden is laid upon our farmers. With the total of American exports annually in the neighborhood of $700,000,000 of which 73 per cent. consists of agricultural products, the force of these observations is at once apparent. We have had many examples of the alertness with which European governments seize every excuse for excluding, or embarrassing American export trade, while England, that most astute of all governments, is putting forth herculean efforts to develop the resources of its own colonies, and thus greatly increase the tremendous competition against which our farmers have to contend. It may not be amiss to call attention also to the deplorable effect of this general system of adulteration upon the morals of our people. Nothing has been, in recent years, more startling than the fact, elicited by inquiry and investigation into food adulterations, that men standing well commercially and socially, who would scorn apparently to do a dishonest action, frequently misbrand their productions, selling articles of food branded as pure, which they know to be impure. This fact was prominently brought out during the lard investigation, when the heads of wealthy and reputable firms unhesitatingly testified to the fact that they sold compound articles of food branded pure.

Other reputable firms, while disapproving of such methods, admitted the practice of them, claiming that unless they adopted the methods of the trade they would be driven out of business ; and one of the strangest sights during the investigation referred to, was that of a dealer praying Congress to enact laws which would compel him and his fellow-dealers to do an honest business. We find that in other countries, notably in England, all the power of the law is invoked to prevent such practices, while in many of them laws have been enacted, directed against the debased hog products manufactured in this country.

The lard investigation brought into prominence another product, namely, cotton-seed oil, as a comparatively new vegetable fat for culinary purposes. Most authorities assert this product to be perfectly harmless, while some few declare it to be injurious. However that may be, it is shown to have been used extensively as a means of adulterating both lard and butter, while it is probably sold as olive-oil more extensively in this country than the product of the olive itself. It is now to be found on the market as a pure vegetable fat, and should its use substantiate what is claimed for it, there is no reason why it should not become one of the most valuable products of the cotton-plant.

COST OF ADULTERATION.

The total value of the food supply consumed in the United States, according to the *American Grocer*, is, at a low estimate, $4,500,000,000. The *Grocer* estimates that 2 per cent. of this, is adulterated, or $90,000,000, of which 90 per cent. is of a character non-prejudicial to health (this is exclusive of meats and milk). Accept for the sake of argument this statement, and the result shows that there is $9,000,000 worth of poisonous food products put annually on the country, and $81,000,000 of fraudulent products. This immense sum of money is simply stolen from the people each year by men who coin fortunes by cheating the consumers.

That this estimate is far below the truth none can doubt for a moment, after an examination of the very able reports made in those States that have attempted to protect their citizens against this crime of adulteration. Take the very conservative estimate of Dr. Abbott, of Massachusetts, of a saving of 5 per cent. to the people in the increased purity of food products (to say nothing of the benefit to health and morals). On $4,500,000,000 the annual saving to the country would be the immense sum of $225,000,000. There is undoubtedly a large part of the food products that never leaves the hand of the producer, and, of course, this is not adulterated; and again the wheat and corn production is rarely found adulterated in this country, besides which there are, of course, many articles manipulated and sold by honest men who would disdain to sell their goods if debased or misbranded. Yet, in spite of all this, undoubtedly the percentage of adulteration, sophistication, and misbranding, largely exceeds, in my opinion, 5 per cent. of the whole, and I am confident that 15 per cent. would be much nearer the mark. Such an estimate would give the startling figures of loss to the people of this country alone of $675,000,000 a year.

Congressional investigation on the hog product of lard clearly shows that my estimate is rather below than above the mark.

THE DEPARTMENT WORK.

The Department of Agriculture has been for a number of years examining into the question of adulterations, and the chief of the chemical division has prepared an extensive report upon this subject, known as Bulletin 13 of his division, of which Parts 1, 2, 3, 4, and 5 are already in print and deserve more attention than a simple reference. It would be well for Congressmen to carefully examine these reports, and they will see that much has been done by the Department in the cause of honest manufacture and in exposing the practices of the adulterators.

In the exposure of frauds undertaken in Parts 2, 3, and 4 of Bulletin 13, no attempt is made to prove that all adulterations are hurtful and injurious to health, as many of them are not. The great trouble lies in the fraudulent practice of selling a cheapened article under a false name and at the higher price of a pure article, thus robbing the producer of the non-adulterated article, as well as the consumer who is made to pay for what he does not want. Doubtless some of these adulterations are due to thoughtlessness and ignorance.

THE EXTENT OF ADULTERATION.

Every article of food is to a greater or less extent the subject of adulteration. The people have no idea of the extent to which this damaging imposition is practiced; from the cheapest and most simple article of diet to the most expensive the art of the manipulator has been applied. The inventive genius of the "American" has been exerted to increase the food supply, undoubtedly a most laudable and praiseworthy undertaking; but when in so doing the manufacturer goes further, and in selling you pepper forgets to add to the words "Pure Pepper," on the label, the words "compounded with burnt meal, mustard, cayenne, buckwheat hulls, pepper dust, etc.," and also tries to obtain the price of a pure article, he degenerates into a common rogue.

Experience proves that a very large majority of the harmless adulterations are used to increase bulk or weight, cheapen the article, and rob the consumer. It is gratifying to know, however, that our manufacturers of food supplies are conscientious enough (in most cases) to abstain from poisoning, even if they do steal.

In the course of this investigation I have visited Boston, New York, Philadelphia, and Baltimore, and have met with a number of gentlemen, who promptly responded to my request for information, and were quite ready to aid in so important a work as the exposure of the frauds perpetrated against the health and commerce of the country. By correspondence and in personal interviews I have failed to find a single, uninterested, person who has not added testimony as to the extent of adulterations, denouncing them as an outrage against the public health and the welfare of trade.

Such unanimity of sentiment, added to the able and voluminous reports of the officers of those States which have undertaken to suppress the nefarious practice, proves beyond question, (1) That adulterations exist to an extent that threatens every species of food supply. (2) That while these adulterations are mainly commercial frauds, practiced by unscrupulous manufacturers, manipulators, and dealers for the purpose of deceiving their customers and adding to their gains, yet there are also, to an alarming extent, poisonous adulterations that have, in many cases, not only impaired the health of the consumer, but frequently caused death.

NEED FOR NATIONAL LEGISLATION.

All of the State officials with whom I have consulted, or whose works I have read, unite in asserting that State legislation alone, can not prevent adulteration, sophistication, and misbranding of the food supply, and the evils that exist will continue until the Federal Government exerts its powerful influence to put down these frauds, alike injurious to health, morals, and fair trade.

The writer, desiring to present the case in as clear a light as possible, and within reasonable limits, has selected from the mass of evidence before him that only which bears the unmistakable stamp of reliability and authenticity, and not nearly all of that. For the greater convenience of the reader, the work is divided into various parts, as follows:

(1) A notice of the work done by the Department of Agriculture.

(2) Extracts from reports of various State authorities and sanitary bodies.

(3) The necessity for inspection of animals intended for food both before and after slaughter.

(4) Public opinion. Extracts, etc., together with a list of adulterants used as far as obtainable.

(5) A list of anti-adulteration laws and where to be found.

Unless Congress takes steps to remedy this evil of debasing and adulterating our food and other products our export trade is certain to be seriously and disastrously affected, and among the many important matters of legislation, that the next Congress will be required to act upon, there will be none more important than that of legislation for the prevention of food adulteration and the misbranding of food products.

The bills reported to the Fiftieth Congress by the Agricultural Committee cover to a great extent the question of live-stock investigation, and as this question has been coupled with the other, I submit a paper on this subject which forcibly presents the necessity for this inspection.

No legislation, unsupported by public opinion, is really effective, hence I have collected and submit as fair an expression of public opinion upon this question as space and time would permit. Such expressions will be found in Part 4.

During the process of the work confided to me, I have necessarily accumulated a mass of evidence from the principal sources already indicated, which it is unnecessary and perhaps useless to include within the limits of this report. It has been necessary to confine myself to that which in my judgment seemed to be sufficient to clearly prove the widespread existence and the dangerous tendency, from a sanitary, commercial, and also from a moral point of view, of this adulteration and debasing of the food and drink supply of the entire country, as well as of medical and remedial agents. The writer, however, is impelled to state before presenting his conclusions, from the evidence in his possession, that it has been a matter of no little difficulty to confine himself strictly, in making this report, to the judicial and impartial terms which its character seemed to require. It would be quite impossible, for the most conservative and moderate of men, to go through the evidence perused by him during the past few months, without being frequently provoked to almost uncontrollable expressions of reprobation, at the overwhelming evidence of fraud, on the part of persons enjoying no little degree of social and commercial prestige among their neighbors. A reckless disregard of the laws of health, and of the moral law seems to characterize a large number of manufacturers of and dealers in products which of all others ought to be pure and nutritious.

Such evidence as the writer presents in the other portions of this report are, in his opinion, amply sufficient to justify the following conclusions:

That, in view of the extensive and increasing adulteration, misbranding, and debasing of food, liquors, and drugs, and in view of the fact that such practices can not be entirely and effectually regulated by State laws, owing to the numerous complications arising from interstate commerce, it becomes, therefore, necessary, that State laws should be supplemented by national law on this subject.

That such national legislation is demanded not only by the State authorities, but by public opinion.

That the consumers of our food products peremptorily, and very justly, demand absolute protection from the evil practices of many manufacturers and dealers, and that wherever the State law can not give such absolute protection, adequate national legislation should be provided.

That, in addition to the interstate commerce feature which must necessarily be regulated by national legislation, it is well understood that no other than national legislation can meet the exigencies of our export trade, which, without the protection of efficient national laws, will certainly and rapidly dwindle into insignificance in the face of a growing and most unfortunate reputation, which obtains in foreign countries in regard to our food products, owing to the prevalence of the practices referred to and the want of an adequate system of national inspection of all goods intended for export.

To enumerate all those to whom the writer is indebted for many courtesies and practical aid in the pursuance of his work would be impossible, but he can not close his report without special mention of Professor Sharpless and Mr. W. W. Kimball, of Boston, and the Hon. F. B. Thurber, of New York, whose public-spirited interest in this work has been as unremitting as it has been valuable.

Respectfully submitted,

ALEX. J. WEDDERBURN,
Special Agent.

REPORTS OF THE DEPARTMENT OF AGRICULTURE ON FOOD ADULTERATIONS.

The chemist of the Department, Dr. H. W. Wiley, is publishing in Bulletin No. 13 an exhaustive study of food adulterations and methods for their detection.

Five parts of this bulletin have already been issued, viz: Part first, Dairy products; part second, Spices and condiments; part third, Fermented liquors; part fourth, Lard and lard adulterants, and part fifth, Baking powders.

Other parts are in preparation, viz, Tea, coffee and chocolate; Sugar, molasses, honey and condiments; Flour and bread; Canned goods, etc. These bulletins can be obtained by application to the Secretary of Agriculture.

STATE AND MUNICIPAL REPORTS.

The people of the country have good reason to congratulate them selves upon the character and ability of the gentlemen who have been selected as food inspectors and commissioners in the few States that have enacted laws regulating the sale and manufacture of adulterated and misbranded articles of diet.

When we consider the small salaries paid these officials, and the tremendous wealth arrayed against them, and the small amount of money placed at their disposal, it is truly remarkable that they have achieved such good results. Nearly every case undertaken against the fraudulent manipulators of food products has ended in a triumph for honesty and a vindication of the laws. Such clear cases of adulteration, misbranding, and selling of diseased and putrid meat have been brought before our courts in Massachusetts, New York, New Jersey, and Ohio as plainly show the necessity for national and State legislation to protect the people of the country, not alone from gigantic commercial frauds, but to preserve the health of our population and enable honest dealers to live while doing a legitimate and honest business. It is utterly useless for the writer to compile a report containing all the matter accumulated by him in the course of his investigation, and were he to attempt it the result would be a publication entirely too bulky for the consideration of the committee of Congress to whom it is to be submitted.

The extracts herewith submitted are presented merely as a few of many similar facts and arguments in favor of national legislation, necessary to protect an honest man in doing an honest business, and to enable the consumer to secure exactly what he asks for, pays for, and has a right to get.

MASSACHUSETTS.

The State of Massachusetts has very carefully investigated the question of food adulterations, and has probably the fullest and most complete laws upon the subject in this country. I glean from the "Manual for the use of Boards of Health of Massachusetts," that there are not only general laws looking to the prevention of adulteration, sophistication, and misbranding and the sale of diseased, tainted, and under-age meats, but special laws relating to the sale of oleomargarine and other butter and cheese imitations, vinegar, milk, and lard. The general laws relate to the adulteration and sophistication of drugs, food, liquor, and provide for the inspection of meats. They provide for fines and imprisonments, but also rely upon publication of the misdeed as a strong preventive. The bill in many of its provisions is similar to House bill 11266, introduced by Mr. Laird in the last Congress and reported favoraby from the Committee on Agriculture.

Dr. S. W. Abbott, secretary of the State board of health, in the sixth annual report of the Massachusetts board of health makes the following statement:

The principal articles liable to adulteration are milk, butter, spices, vinegar, cream of tartar, and various sorts of drugs. The value of these articles specified, at present consumed in the State, may be stated in round numbers at $15,000,000 annually.

It may safely be stated that the enforcement of the statutes has resulted in a saving to the consumers of at least 5 per cent. of this amount, or $750,000, a sum equal to seventy-five times the amount expended in the enforcement of the laws.

REPORT ON ADULTERATION.

The result of Dr. Wood's analysis for 1886, as reported to the Massachusetts board of health, pages 109, 110, 111, show the following adulterations:

Coffee essences.—Two samples, consisting largely of chicory, burnt starch, and caramel.

Molasses.—Sixty-eight samples examined with special reference to presence of tin. In 33 samples tin was found.

Honey.—Out of 7 samples 5 were adulterated with cane sugar and glucose.

Baking powders.—Five samples, analyzed for alum and all found to contain it.

Cream of tartar.—Eighty-four samples examined, 64 of which contained less than 6 per cent. of impurity. Twenty were adulterated with flour, terra alba, or other foreign substances varying in extent from 10 to 82 per cent.

Mustard.—Forty-six samples, 20 genuine, 26 adulterated.

White pepper.—Fourteen samples examined, of which only 7 were genuine. The adulterants were wheat, rice, and ginger.

Black pepper.—Sixty-two samples examined, 37 genuine and 25 adulterated. The adulterants used were chiefly wheat, rice, and corn.

Cayenne.—Four examined, 2 adulterated.

Mace.—Two samples examined, both adulterated.

Cassia.—Examined 13, all genuine but 1.

Cloves.—Examined 12, all of fair quality.

Ginger.—Examined 54, of which 45 were genuine and 9 adulterated. The adulterants were wheat, rice, and corn.

TIN POISON IN MOLASSES.

Thirty-three cases of adulteration of molasses *with salts of tin* are cited in Dr. Wood's report for 1886. Speaking of this article Dr. Abbott says, page 84:

It was also found by experiment that the same salt was exceedingly poisonous when administered to animals in *moderate doses,* one-fourth to one-half grain producing symptoms of acute poisoning, *followed by death in a few minutes,* the drug being administered in various modes, both in solid form and in solution. The lesion of internal organs caused by the ingestion of the poison were well marked, the stomach especially presenting the appearance caused by an active corrosive poison. The substance was also detected in the liver.

Of 1,468 samples of food (including milk) and drugs examined in Massachusetts for the four months ending September 30, 1886, 926 were found good quality, and 542 not conforming to statutes. It will be found from this that even after the continued enforcement of the anti-adulteration laws from 1882, when food laws were just enacted, that there still remained in 1886 over one-third of the products examined "not conforming to the statutes."

RELATIVE TO CALVES.

It is unlawful in Massachusetts to kill or sell any calf under four weeks old, and when such veal is found it must be destroyed. The penalty for knowingly selling such a calf is imprisonment not to exceed six months and a fine not to exceed $200, or both.

NEW YORK.

I am greatly indebted to Mr. Lewis Balch, secretary of the State board of health, for their valuable reports and the promptness and courtesy with which they were sent. These reports extend through a series of years, and are not only comprehensive but exhaustive.

I find the following in the reports (on page 20, report of 1888):

The analyst of food has examined a considerable number of largely-employed food articles and accessories and has pointed out various fraudulent practices which should be suppressed. Fortunately most of the common sophistications are such as to affect the pocket of the consumer more than they injure the health, but they are none the less to be condemned.

From the same report we take the following extracts as showing the extent of adulterations in food and drugs in the State of New York, as shown by the highest official authority. The report of Prof. S. A. Lattimore, public analyst, State board of health, New York, shows the result of 60 analyses, 14 of which were pure and 46 adulterated, the adulterations being—

Black pepper 3
White pepper 2
Cream of tartar 41

The conclusions reached by the professor are worthy of note. He says on page 60, report 1888:

In the great majority of cases it has been found that care and intelligence are exercised, and as the amount of capital required to maintain such a business with profit is very considerable, the proprietors fully appreciate the importance of securing a high reputation for the excellence and purity of their products. *The fact that the State exercises a supervision over such matters is doubtless not without silent but salutary influence.* *

It is, however, in the minor food articles, which are far more costly, but are consumed in smaller quantities, that adulterations are chiefly practiced. The chief reasons for this, besides the relatively high cost of the raw material, are, that as the necessary machinery is inexpensive the manufacture is often carried on in a small way, and the very common practice of selling such goods to retail dealers in bulk destroys all means of tracing them to the manufacturer, who makes no effort to establish a reputation, and who even attempts to palliate his dishonest practice by the claim that the substitutes which he employs are harmless, and that he *is compelled to pursue such a course because his competitors do.* *

Of 48 analyses made by Professor Lattimore of useful household articles only 7 were pure, the remaining 41 containing starch, terra alba, superphosphate of lime, in combination with cream of tartar, while 30 of the total contained *not a particle of cream of tartar,* but consisted of tartaric acid and starch, terra alba, superphosphate of lime, alum, and starch, in various combinations.

Regarding spices, this gentleman says:

While many manufacturers send out under their own names ground spices free from all foreign substances and of excellent quality, and find an increasing appreciation of their goods, yet the markets are still flooded with articles of poor quality, originally and largely mixed with any convenient rubbish which can be manipulated into the semblance of the genuine article. Fortunately for the victimized purchaser, the substitutes by the dishonest spice-grinder, however unsuited for food, and often repulsive in character, are not positively poisonous.

CANNED PEAS AND BEANS.

The principal complaint in these articles is that sulphate of copper has been used so extensively for the purpose of greening that they are exceedingly dangerous. As far as I can learn the German, French, and English Governments do not prohibit the use of this article for this purpose. Prof. R. Ogden Doremus, M. D., LL. D., professor of chemistry and toxicology, Bellevue Hospital, New York, in writing to Mr. Jas. P. Smith, says that "this method of greening is not injurious to health," and adds that Pasteur, Brouardel, Galippe, Gauthier,

---

* The italics are mine.

Proust, Gallard, Honnerkopf, Stubenrauch, Rademacher, Muller von Pforzheim, and others after careful investigation proved that the small amount of copper used in canned vegetables is not hurtful. (See page 410 of New York State Board of Health for 1887.)

RECOMMENDING FEDERAL LAWS.

Frederick Carman, assistant secretary of the board of health, on page 419, report 1887, makes the following recommendations:

(1) The laws should be amended so as to define the normal constituents of malt beers and malt beverages.

(2) That a malt beverage should, as its name indicates, contain only malt and hops with their constituents and water.

(3) That Congress should pass a law making it obligatory upon the Internal Revenue Department at all times to permit publicity to be given to the ingredients reputed to it by the brewers and distillers as having been purchased from which to brew or distill.

VINEGAR.

In 1886 the public analyst examined 74 samples of vinegar: 14.8 per cent. contained 5 per cent. or over of absolute acetic acid, and 63, or 85.2 per cent., contained less than 5 per cent., and therefore fell below the legal requirement.

There are numerous vinegars which may be equally as good as cider vinegar, but, in the language of the New York report, page 371—

While vinegar made from spirits or other wine may be equally wholesome, it ought not to be sold for cider vinegar, as is very frequently the case. An article so largely used in the preparation of food as vinegar ought to be both free from adulteration and of good strength as well, but the results of the examinations so far made show that here as elsewhere wide differences in quality exist. The addition of mineral acids is very uncommon, but much vinegar is sold which has been plentifully watered, and the greater part of that sold as cider vinegar is a so-called white-wine vinegar, colored by caramel, with perhaps some cider vinegar added to give it flavor.

The New York law requires a standard of 4.5 per cent. of acetic acid in vinegar.

OTHER FOOD PRODUCTS.

[See page 429 in report of Dr. Lattimore for 1887.]

Of 376 articles of diet in daily use in every household 255, or more than two-thirds, were adulterated. Of 205 samples of so-called cream of tartar analyzed, only 53 were unadulterated. Among the adulterated samples were found oxalic acid and terra alba (white earth), terra alba and starch. The quantity of this poisonous acid was about 5 per cent. In referring to manipulated spices, Dr. Lattimore says (page 425):

The articles used for the purposes of adulterations are extremely numerous. Most farinaceous substances which have become damaged and unsalable may by skillful roasting and grinding be rendered serviceable by the "spice mixer." Many other articles which might be included under the general rubbish, by suitable manipulation may be transformed into mixtures which closely resemble the various spices in color and appearance, lacking only a little seasoning with the smallest possible quantity of the real article to give the characteristic odor and fit them for the market.

It would be interesting, did space permit, to go more fully into detail so as to show a comparison of various adulterations after the enactment of the anti-adulteration statutes in New York State, but this can not be done here, and my observations lead me to conclude that the failure to enact Federal laws render the local laws to a great extent ineffectual, and inoperative.

NECESSITY FOR PUBLICATION.

The failure to make public all adulterations makes the practice more common than under a system by which the manipulator understood that his practices would be thoroughly published.

DRUGS.

Prof. W. G. Tucker, in his report (page 250, report 1888) shows the result of 326 samples of various drugs, as follows:

Good quality .... 140, or 43. 0 per cent.
Fair quality....................................:.................. 44, or 13. 3 per cent.
Inferior.. 79, or 24. 2 per cent.
Not as called for.. 63, or 19. 3 per cent.

The latter 63 samples, designated "not as called for," show simply the fairness of the examiner, as, while containing adulterants, benefit of the doubt is given to the compounder or seller, whether the article was sold through ignorance or mistake. The fact established is that out of 326 samples examined only 140 were pure, and 79 came under the heading of "inferior," which the writer says is used in the report to designate articles "if not clearly adulterated or falsified, lacking in any important constituent, deficient in strength from improper manufacture, partial or complete decomposition, or other causes, or containing undue amount of impurity."

It was shown that in the purchase of samples great care was taken in each and every instance, a written order was tendered, which order gave in full the official names of the articles called for and the amounts desired.

In closing his report Professor Tucker says:

That the work done during the past two years has had a decided effect in improving the quality of drugs sold throughout the State, there can be no doubt, and if manufacturers, dealers, and others interested will co-operate with the board in the efforts it is making to raise the standard of the drugs and medical preparations now upon the market, the desired end can be obtained without interfering with any legitimate business industry, or embarrassing dealers by a too rigid or oppressive enforcement of the law.

QUININE, MORPHINE, AND IRON.

In the report to board of health, New York, October 31, 1887, by Prof. G. C. Caldwell, speaking of the assays made in morphine, quinine, and iron pills, he says:

By the methods used, giving every fair advantage to the dealer, few of the morphine pills came up to the standard.

There was an improvement in the standard of quinine pills. The citrate of iron and quinine is often below the standard and the quality of the sulphate is usually poor. Druggists who have been warned to cease selling certain brands of alkaloidal preparations found to be seriously deficient in quality, complain that they do not know where to go for their supplies, especially since houses of wide reputation not infrequently send out such deficient goods, and it appears to them that the board of health should go at once to the wholesale dealer with its warnings.

The result of the analyses made by Professor Caldwell show : Sulphate of quinine, 14 samples, only 2 of which did not contain foreign substances ; citrate of quinine and iron, 16 samples, of which only 3 were free from adulterants ; pills of sulphate of quinine, 58 samples, 28 of which contained more or less adulteration ; pills of sulphate of morphine, 28 samples, not one of which was up to the standard claimed.

COMMERCIAL AND HEALTH FRAUDS IN BROOKLYN.

The result of the laboratory work of the Brooklyn Bureau of Health offers conclusive evidence that the position taken elsewhere, that food products are adulterated much more largely than 5 per cent. is correct. In presenting a partial table of the results obtained, it is right to quote what the Commissioner says (page 66) :

There are large numbers of adulterations practiced with little if any attempt at secrecy which we have not been called upon to investigate.

Had these commercial frauds been examined doubtless the per centage shown would have been much greater.

[Pages 69 and 70 of the report for 1885.]

No. of samples.	Articles.	Adulterations.
95	Preserved fruits and jellies	94 with tin, 34 with copper.
1	Tomatoes	4.5 mgm. tin, 13.5 mgm. lead, 30 mgm. copper.
2	......do	Both tin and copper.
1	Gelatine	Poor quality.
1	Ice cream	With gelatine.
24	Milk	4 watered.
7	Butter	Genuine.
1'	Condensed cream	Not good.
2	Evaporated milk	Skimmed before evaporated,
7	Condensed milk	5 partly skimmed.
2	Pickles	1.8 mgms. copper and 1 with 14 mgms. copper.
4	......do	3 copper.
1	Whisky	Tannic acid.
3	Beer	All zinc and copper.
3	Ale	All copper, lead, and zinc.
1	Cocoanut candy	Adulterated.
92	Candy	18 colored with lead chromate.
18	Candy colors	1 lead chromate.
7	Cream of tartar	Impure.
2	Pepper	1 adulterated.
3	Rough on rats	Consisted mostly of arsenic.
8	Soda water	2 contained traces of copper.
1	Fruit syrup	Colored with cardinal.
1	Lemon syrup from metal tank	Contained tin and antimony.

Some artificial jellies have been sold in Brooklyn which are made from gelatine, sugar, artificial flavoring extracts, cochineal, or analine red.

The following articles were found in products purchased of Brooklyn Polishing Companies :

French chalk (white) also soapstone or talc, is a silicate of magnesia.
Gum arabic.
Charcoal, from wood.
Drop black, animal charcoal, or burnt bones.
Yellow ochre. A mixture of clay and hydrated oxide of iron.
Venetian red. A red earth, mostly oxide of iron and clay.
Chrome orange. Chromate of lead, mixed earth, yellow ochre. Poisonous.
Burnt umber. Differs very little from ochre in composition.
Silesia blue. A mixture of ochre and Prussian blue. Gives off cyanagen abundantly in heating.
Turmeric. The powdered root of *curcuma longa.*
Indigo and indigo extract. (This latter is a sulphate of indigo and indigo carmine.)
Persian berries, with ochre, giving a grayish-green colored pigment.

Of these the Commissioner (page 121) says *chrome orange* and *silesia blue* are poisonous. Of the remaining colors he says :

While not so injurious, they are used to deceive the public, and should be regarded as adulterations.

On the examination of twenty samples of coffee, polished, purchased on the open market, lead was found to adhere to the bean in considerable quantities, and the use of lead and Prussian blue is shown to be common.

Alum and sal-soda are added to the Persian berries to make a beautiful lake color. This separated, and mixed with indigo gives an olive green.

In addition to this system of polishing, I learn that coffee is so manipulated as to extract the essential oils and essences from the bean, which is sold as coffee, while the essences are manufactured in extracts, and sold as such.

Another method of adulteration is to take the shriveled beans and soak them in sea-water, which "plumps" the berry and deceives the public into buying a poor article of coffee for a good one.

The natural sequence is that if you desire a fine grade of coffee you will avoid the highly polished, colored and plumped grains, and purchase the poorest looking coffee you can find.

BOLOGNA SAUSAGE COLOR.

Color is an essential in selling many of the articles of food offered upon the market and we not only have "butter color" and milk color, but copper in various ways is used to "green" beans, peas, and pickles, but the Brooklyn health department has discovered a coloring for Bologna sausage composed of salt, saltpeter, borax, alum, and Venetian red. The harmfulness of this mixture consists in its ability to deceive the purchaser into purchasing stale, partially decomposed, or tainted meats, which should cause its use to be prohibited.

TIN AND LEAD IN CANNED GOODS.

The Brooklyn chemist finds the presence of lead and tin in large quantities in unsealed cans and tin pails, while in hermetically sealed cans the tin and lead does not seem to be affected to any considerable extent when deprived of the air, but as shown here the case is quite different. (See report 1885, page 134.)

COLORING MATTER IN FOOD.

The vermacelli manufacturers not only use poor flour and whiten it with pipe-clay or kaoline, but the absence of eggs is supplied by saffron, tumeric, and chromate of lead. The practice of the use of chrome yellow has decreased materially since inspection has given publicity to the practice in New York State. Aniline red or *fuchsin* was found to be used in coloring head cheese.

Speaking of these practices the Brooklyn report says (p. 132):

The manufacturer colors his meats to please a public taste for a bright red color, but when the public learn that such meat may be colored with harmful colors, while the meat itself is the worst salable meat in the market, the manufacturer will not be called upon to practice it. Aniline red is in itself an objectional color, because many samples in the market contain arsenic.

The supposably wholesome and toothsome licorice drops given the children for colds are said to be made from the candy factory sweepings, and are apt to contain a good deal of foreign substances as well as dirt. Several cases of sickness have been reported from their consumption.

Pickle greening is a source of poisoning that all know of, and few pay any regard to the addition of copperas. All the American cater seems to care for, is that the article be pleasant to the eye and pleasant to the palate; whether it be *wholesome* is not taken into consideration.

The Brooklyn health commission says the proper dose of sulphate of copper is from one quarter to 2 grains, while in "greened" pickles analyzed three small pickles contained the full medicinal dose and 5 grains was contained in one-quarter pound. On page 110 an interesting case is cited, that of Maggie Martin, a little girl, who was killed by eating a part of one large pickle greened with sulpate of copper.

Dr. E. H. Bartley, chief chemist of the Brooklyn health department, in his report for 1887, speaking of the result of the inspection of meats, vegetables, and other food products, says:

Although it can not be claimed that all improper articles of food have been detected it is certain that fewer complaints of sickness having resulted from various kinds of foods have reached us during the past year than any year since 1883.

The cause of this decrease the doctor ascribes not only to the work of his department, but to the fact of the education of the people to seek better goods, which induced dealers to comply with the demand. He adds that he is convinced that the responsibility rests upon the people who are so ready to protest against adulteration. He says:

There can be no other reason given for the wholesale use of colors in the preparation of confectionery, pastry, ice-cream, green peas, pickles, macaroni, milk, butter, coffee, tea, etc., than that the public demand them. (Dr. Bartley's report page 33, 1887.)

Not less than ten or twelve of these colors used in these various foods have been condemned and their use prohibited. During the past few months one large confectionery factory was found to be using two poisonous colors. One macaroni factory in this city was found to be using Martin's yellow or triniter napthaline, to impart a rich color, in imitation of the color imparted by eggs. The same poisonous color was found in another factory, but there was no evidence that it had been used. This color is sold under the name of artificial saffron and other trade names to conceal its identity. The demand for these yellow-colored foods has repeatedly led to the use of chrome yellow and other poisonous colors.*

Dr. Bartley pertinently remarks:

It is not enough to say that the amount of these poisonous colors used is too small to do harm. Any amount of a poison is too much to be allowed to be used for human food, especially when it is to be administered by an ignorant servant or careless workman.

He cites a case of a family receiving a sample of pistachio ice-cream by Bretnis Green, and flavored with bitter almonds. On the subject of the poisonous creams Dr. Bartley says (page 34):

It seems difficult to suppose that an ordinary palate or stomach could be deluded into the idea that it was receiving strawberry ice-cream because the latter contained some red aniline and amyl ethers. A very small amount of fusel oil (amylic alcohol) in a whisky is popularly regarded as very detrimental to the health of the consumers, while a plate of some kinds of ice-cream or a glass of soda-water frequently contains more fusel oil than five glasses of poor whisky.

Speaking of soda, an interesting paper can be found on pages 51 and 52 of the New Jersey dairy commissioner's report for 1888, in which analysis showed the presence of lead, copper, zinc, and tin. The commissioner recommends that the use of bottles whose stoppers consist of "a loop of coated brass wire and a tin button containing lead with a rubber belt be prevented."

Dr. Bartley, in referring to the coloring of peas, beans, etc., with copper, says the board of health had legalized them under certain conditions, viz, that copper not to exceed three-quarters of a grain to the avoirdupois pound of peas or beans, equivalent to 3 grains of crystallized sulphate of copper, and that it should be plainly stated on the label. The regulations of the board, both as to labeling and amount of copper used, it seems were frequently violated, even by the very parties who instigated the regulations. I am glad to add that Dr. Bartley secured the conviction of two parties brought to trial for violation of the rules of the board.

[When it is recollected that these colored and poisonous vegetables are brought into direct competition with the product of our market gardeners, and not only

* See Report of Philadelphia Board of Health, referred to elsewhere.

depreciate the price of our home product, but spread broadcast disease, and further, that they come principally from France, a country that excludes the hog product of our farmer, it seems reasonable to suppose that Congress would not hesitate to exclude the importation of these poisonous products.—NOTE BY THE SPECIAL AGENT.]

NEW JERSEY.

The report of the dairy commissioner of New Jersey for 1888 contains much valuable information. Mr. William K. Newton, the commissioner, has prosecuted his work with a zeal and energy which deserves the thanks of the people of the whole country. His exposure of frauds can not fail to convince every intelligent reader of the necessity for stringent legislation. According to his report $1,100 have been paid into the State treasury in fines for violation of the oleo laws, while the penalties paid for violation of the milk laws have reached $550.

FOODS.

The following table of foods analyzed by the commissioner, with the results of such analysis, tells the story briefly but pointedly:

Articles of food Examined.

Article.	Total.	Pure or standard.	Adulterated or not standard.
Butter and oleomargarine	68	24	44
Milk	121	78	43
Lard	141	80	61
American canned goods	55	54	1
Imported canned goods	16	5	11
Ground coffee	24	8	16
Coffee essence	1		1
Tea	10	10	
Black pepper	28	6	22
White pepper	4	2	2
Cayenne pepper	3	3	
Mustard	41	1	40
Cinnamon	20	7	13
Cloves	13	3	10
Allspice	8	2	6
Ginger	11	8	3
Nutmeg	2	2	
Extracted honey	10	5	5
Maple syrup	9	5	4
Molasses	2	2	
Drips	1		1
Vinegar	12	1	11
Pickles	7	5	2
Carbonated beverages	7		7
Total	623	320	303
Per cent		51.36	48.64

The commissioner says:

While the adulterants used are not surely harmful to health, they are always fraudulent, and deprive the purchaser of an amount of money equal to the amount of adulteration; hence if the law can but restore to the people an amount equal to that of which they are defrauded by the adulteration, many times the sum appropriated for this work shall be returned to the pockets of our citizens.

Four hundred and fifteen articles of drugs were analyzed; of these nearly 50 per cent. were adulterated.

PURE LEAF LARD.

The commissioner says that the probable amount of pure leaf lard prepared in New Jersey would not reach 1,000 pounds.

TEA.

Of tea he observes that:

The results obtained show that, while there is no adulteration, there is large amount of inferior and debased tea sold in the State. The United States law prevents the importation of the adulterated article, but much that comes to this country is of very poor quality, having little of the flavor of the true leaf.

SPICES.

On the subject of spices he says:

The following facts are deduced from the results of the investigations : (1) A few dealers put up nothing but absolutely pure spices, and every package examined bearng those dealers' names proved to be of excellent quality. (2) Several spice-mills and wholesale dealers put two grades on the market—one pure the other adulterated. (3) Many houses selling spices sell nothing but adulterated articles. (4) Spices sold in bulk are almost invariably adulterated. (5) The price paid the retailer is no guaranty of the quality. (6) The retail dealer and the consumer are jointly responsible for the adulteration—the former by encouraging the sale of cheap and impure articles, the latter in trying to get spices at a less price than will warrant the sale of pure articles.

BAKING POWDERS.

On this subject he says:

The greater number of the inferior grades of bakingpow ders were only to be had at stores patronized by poor people, or those of moderate means, hence we are forced to believe that any fraud in these articles is perpetrated at the expense of those least able to bear it.*

A baking powder should answer to all of the following requirements: (1) It should generate the maximum amount of gas. (2) It should contain no unhealthful ingredients. (3) It should leave in the loaf no unhealthful residue. (4) The elements should be combined in such proportions that the residue is neutral in reaction.†

BRANDY AND WHISKY.

Of 43 samples of brandy analyzed 6 only answered to the tests required by the United States Pharmacopea, and 37 were inferior. The report says:

The difficulty of obtaining pure brandy of a proper age for medicinal use is very great. This is especially true of the imported article, while that made in California is, as a rule, of inferior quality and not sufficiently aged and bland to be used in cases of illness. The following statement, made in the United States Consular Reports, November, 1887, page 333, is interesting in this connection:

"The term 'brandy' seems to be no longer applied to a spirit produced by the fermentation of grapes, but to a complex mixture, the alcohol of which is derived from grain, potatoes, or, worst of all, the refuse of the beet-root refineries. It would seem to be fairly impossible at present to purchase a pure cognac, as each individual pro-

---

25

prietor of a vineyard has become a distiller and compounder. He has acquired the art of imitating any special flavor or vintage of brandy that may be called for. Potato spirits and beet alcohols, the most deleterious and obnoxious of all the varieties of spirits, are sent from Germany into France in vast quantities. They are flavored, colored, and branded or labeled to meet the wishes of American connoisseurs. The mere fact of coming out of bond, or 'straight through the custom-house,' is generally accepted here as sufficient evidence that they are pure and genuine. It is rather unfortunate that physicians themselves frequently strengthen this hallucination in favor of imported spirits by giving the most stringent orders to their patients to procure genuine French cognac, even though it may command tenfold the price of an absolutely pure spirit of domestic production. This imperative command becomes a cruel injustice in the case of poor patients. Under the best of circumstances, what is there to be gained by the use of French brandy in preference to pure domestic spirit?"

And, it may be added to this statement, if alcoholic stimulants are to be prescribed by the physician, let him first ascertain the source of the sample and acquaint himself with the quality, origin, and ingredients.

Of 15 samples of whisky, 3 only were equal to the requirements of the pharmaçopœia.

OPIUM AND OTHER DRUGS.

Of 42 samples examined only 8 were standard; as previously stated, other drugs showed 55.66 per cent. of debased and adulterated articles.

OHIO REPORT.

The following extracts taken from the report of the dairy and food commission of Ohio for 1887, speak for themselves and are submitted without comment.

MAPLE SIRUP.

Two fraudulent samples analyzed. One contained 62 per cent. of maple sirup and 28 per cent. of glucose; the other, 25 per cent. maple and 75 per cent. of glucose sirup. One of the parties pleaded guilty and paid $500 fine. The other swore the goods were sold before the law was enacted, and was acquitted.

VINEGAR.

Commissioner Hurst, speaking of this article, says:

The vile and hurtful compound manufactured from drugs and acids should be forbidden a place in the trade, whether made within the State or without. And the "low wine" vinegars should be branded with their true name, and the coloring of such vinegar to simulate cider vinegar should be positively prohibited. Such an amended law would add $100,000 per year to the value of the Ohio apple crop and greatly promote the public health.

BAKING POWDERS.

Thirty samples were examined. The commissioner evidently tried to be impartial, as he issued a circular to the trade before purchasing the samples. His classification of the powders were as follows: First quality, cream of tartar powders; second quality, phosphate powders; third quality, alum powders.

The analysis resulted as follows:

First quality, cream of tartar... 8
Second quality, phosphate... 2
Third quality, alum.. 20

ORANGE CIDER.

An article claimed to be pure orange juice was found on sale as orange cider, and upon analysis proved to be only "sweetened water, sharpened up with citric and tartaric acids and flavored with the oil or extract of orange." The cost was about 16 cents. It was retailed at $2.50 a gallon.

The sale of this product reached very large proportions, but in *ten days after the fraud had been published* the whole business collapsed. (See page 14, Ohio D. Report.)

In many lines of food and drink products, however, the evils can only be ended by national laws and by the establishment of a national commission, which can reach the great manufactories, refineries, rectifying establishments, spice-mills, etc., and have supervision of all the channels of commerce and trade without the obstructions and limitations of State lines.

Mr. John J. Geghan, assistant dairy and food commissioner for Ohio, in his report says:

A great amount of baking powder sold in our market is not fit for food, and should be condemned as injurious to health. There is such a competition among the manufacturers of this so-called baking powder that with each bill of this stuff, which they sell to retail grocers, they give a large amount of glassware, which is offered as an inducement to the purchasers. This is a nefarious scheme and should not be countenanced by the public. But it seems the public likes to be humbugged; reasoning sense ought to teach us that the manufacturer of baking powder can not afford to give glassware as a present to each customer unless he makes up for the cost of the glassware by the high price he receives for his worthless stuff. Baking powder is used extensively, and the ingredients entering into its composition should be pure and healthy. Alum, an injurious mineral, is substituted for cream of tartar, a vegetable substance, so as to cheapen the cost to the manufacturer of the so-called baking powder. If the manufacturer is honest he should discontinue the custom of offering prizes as an inducement to the people to buy his worthless goods, and use in the manufacture of baking powder good and healthy substances.

Mr. Geghan, it will be seen, pronounces alum an unhealthy mineral. Commissioner Hurst, in his circular heretofore alluded to, on page 8 says:

Pure alum is undoubtedly a hurtful salt, and the resultant salts from its combination with soda can scarcely be less hurtful.

And yet this is a question about which "doctors disagree," and any number of conflicting opinions and certificates can be had from eminent chemists on either side of this question.

The official investigation of this class of baking powders made in

England to test their healthfulness resulted in their favor. But this conclusion rested upon the statement of the chemists that the resultant salt of hydrate of aluminum remaining in the bread was insoluble, and hence unhurtful when taken into the stomach. But some of the ablest chemists of this country declare that hydrate of aluminum is quite soluble, and hence is as hurtful as the alum in other forms. So that the question is still an open one to be determined by further careful scientific investigation.

Dr. Hassall says of alum:

"Either alum is very apt to disorder the stomach and to occasion acidity and dyspepsia."

The New Jersey food commission says:

Our investigations show that while especially the higher grades of cream of tartar and acid phosphate of lime powders are maintained at a quite uniform standard of excellence, the State is flooded also with many baking powders of very poor quality—cheap goods, poorly made. Of the thirty-nine brands examined, twenty-five contain alum or its equivalent, in the shape of some soluble alumina compound; eight are cream of tartar powders, with small quantities of other ingredients in several cases; four are acid phosphate of lime powders; two belong properly under none of the above classes.

With one exception, the powders containing alum all fall below the average strength of the cream of tartar powders, and in the majority of cases they fall much below the better grades of the cream of tartar powders.

In the cream of tartar and the acid phosphate of lime powders no indications of substances likely to be injurious to health, in the quantities used, have been found.

There appears to be ample ground for requiring that the makers of baking powders should publish the ingredients used in their powders, in order that the consumer, who may justly have doubts of the desirability of using certain kinds, may be protected. At present the only guaranty of an undoubtedly wholesome and efficient article appears to be the name of the brand.

Referring to Dr. Mallet's experiments, he says:

He regards it as a fair conclusion that not only alum itself, but the residues which its use in baking powder leaves in bread, cannot be viewed as harmless, but must be ranked as objectionable, and should be avoided when the object aimed at is the production of wholesome bread.

These experiments of Professor Mallet are conducted in the right way, and his conclusions are entitled to great weight. We might quote the decided opinions of many scientific men against the use of alum baking powders, but with the preceding facts before us this phase of the question may be left.

DAIRY PRODUCTS.

Of dairy products Mr. Geghan says:

The consumers of oleomargarine—and they consist generally of the industrial or working classes—by using fraudulent articles of food, are compelled to pay 20 cents a pound for a mixture of lard and tallow that can be purchased in their original state for 5 cents a pound. It should be driven out of the markets as an unhealthy article of food. Surely it is more economical for the housewife to pay 25 cents, or even 30 cents, for pure creamery butter than to pay 25 cents a pound for tallow or lard.

There is a principle involved in a manufacture of oleomargarine which should separate it entirely from the question of cheap butter. Let this oleomargarine take the place entirely of butter made wholly from cream, and we immediately curtail our beef supply, an article of food which the American people, more than any other nationality, use so much of, owing to its nutritious quality; because in order to manufacture oleomargarine you must first kill the animal to get in part the necessary ingredients, and thereby create a scarcity of the beef-producing species. If we repudiate the artificial and use only the pure article we encourage the farmer to raise more cows. The more cows he raises the greater will be the supply of beef, pure milk, butter, cream, and cheese. No farmer will undertake to raise or propagate these species of cattle unless he can realize a profit by so doing.

Mr. Geghan estimates the loss to the farmers of the country by the manufacture of oleomargarine to be for 1886 $70,000,000.

In spite of the United States oleomargarine law it would seem that the profits are so great in selling oleomargarine for butter, that unscrupulous men are constantly tempted to sell it, and I am sure that in many of the smaller cities of the country, some not far from the national capital, oleomargarine is openly sold in violation of Federal statute. In confirmation of this idea I find that even in Ohio, with its vigorous State officers, alert and watchful, ready to enforce the State laws and aid in the enforcement of the internal-revenue laws, the provisions of the oleomargarine bill are constantly violated, as well as that of the State laws relative to milk and cheese.

Of 97 analyses of butter made by the chemist—one known to be genuine—50 were adulterated, and only 46 genuine. Of 43 samples of milk analyzed, 7 were known to be pure; of the remaining 36, representing milk sold in Columbus, 18 samples fell below the standard required by law.

The law relative to skim-milk cheese seems to have been made to be broken in Ohio, but the prompt prosecution and conviction of the offenders, who pleaded guilty and paid fines of $50 each, seems to indicate that in the future the condition of cheese and of morals will both be healthier in Ohio.

Full cream cheese in Ohio must mean something; it must mean exactly what it says after this, or you must repeal the present dairy laws and abolish the commission. (Henry Talcott, assistant dairy commissioner for Ohio, page 32, report 87.)

BEER ADULTERANTS.

Dr. Wm. Dickore, of Cincinnati, in a report of certain analyses made by him for the dairy and food commission (page 24), says:

" Beer in its perfect condition is an excellent and healthful beverage, conferring in some measure virtues of water, wine, and of food, as it quenches the thirst, stimulates, cheers, and strengthens." (Andrew Ure.)

The above is the opinion of this great scientist. While most brewers certainly aim to produce beer of the best quality, still there are many others unscrupulous enough to substitute for the sake of cheaper production a very large percentage of other grains for malt, and also other drugs, like quassia, aloes, picric acid, etc., in connection with artificial flavors, to replace the more expensive hops. I would like to state that even well-meaning brewers may make the mistake and use too many hops, which will produce an unwholesome beer. Beer should neither be too strong in alcohol or be too highly hopped. While barley is (by practical experience of centuries) the most appropriate cereal to be used in brewing beer, still other kinds of grain, as rice or corn, in very limited quantities, may be admitted in manufacturing certain kinds of beer. The proportions should however be regulated by law.

As a fraudulent adulteration corn stands in the same line with the adulteration of beer as low wines stand in the adulteration of cider vinegar, and when used to excess is more injurious to the system than any other kind of beer.

WINE.

The assistant food commissioner for Ohio says (page 29):

The grape-growers of Ottawa county have had their industry ruined by the manufacture of bogus wine out of foreign substances, so that the demand for grapes for this purpose no longer exists to any great extent.

PENNSYLVANIA.

The committee on adulterations, poisons, etc., of the Pennsylvania legislature, speaking through their chairman, Dr. Pemberton Dudley (see page 90, report of the Board of Health of Pennsylvania), says:

There can be no question, however, that the department of sanitary labor assigned to this committee is one of the most important that engages the attention of sanitary authorities. The adulterations of food and drugs are so numerous, so common, so universal, we might almost say, and at the same time so prejudicial to the health of our people, that constant watchfulness and omnipresent oversight alone can repress and prevent them.

Wherever competition prevails there we find the temptation to lower the standard of purity and strength of our food-stuffs and our medicinal preparations, and with the exception of the few that are protected by patents, this competition extends to all.

Dr. L. Wolff, in an article on "Our Drugs and Medicines" (Pennsylvania Board of Health report, page 338), says:

The use of pure drugs and medicines, properly compounded and administered, constitutes a most important feature for the preservation of health and the prevention of avoidable death. In all civilized countries it has been made the duty of the state to control and supervise this through competent officials and special laws. The harm arising from inert or impure drugs consists not only in defeating the end and object they are intended for, by admitting of the unchecked progress of the disease and the fatal consequences thereof, but also in their improper and poisonous admixtures, which make them destructive to life and health. Many of them possess powerful and toxic action, and consequently, when compounded and administered in improper quantities and doses, give rise to most disastrous results.

And again:

That there are annually a number of valuable lives sacrificed from this cause is as little to be doubted as that all the cases of suffering, illness, and death therefrom are certainly avoidable by proper knowledge, forethought, precaution, and *legal supervision*.

CHROME YELLOW POISONING.

In a pamphlet issued by the State Board of Health of Pennsylvania, in 1887, entitled a Clinical Analysis of sixty-four cases of poisoning by Lead Chromate (Chrome Yellow), used as a cake-dye and prepared by Dr. D. D. Stewart of Jefferson Medical College of Philadelphia, we find the following conclusion:

There is now no question, from recent developments, that the poisoning has been going on unsuspected in various sections of the city for years. Since my attention has been directed to the matter I have been surprised, in looking over the mortuary records of the past few months, to see the large number of deaths returned as convulsions,

SLAUGHTER-HOUSE AND DAIRY INSPECTION.

I will introduce this division of my report with a paper prepared at my request by Dr. D. E. Salmon, chief of the Bureau of Animal Industry, whose experience and familiarity with the subject especially qualify him to speak of it with authority.

THE NECESSITY OF INSPECTING ANIMALS SLAUGHTERED FOR FOOD.

The inhabitants of the United States are the greatest consumers of meat of any people in the world. They have developed this characteristic because nowhere else has the supply of meat been at the same time so abundant, so cheap, and of such superior quality.

We find, however, that the conditions under which our food-producing animals are grown, marketed, and slaughtered have been rapidly changing, and that this change has greatly increased the desirability of government supervision over the transportation and slaughter of animals and the preparation of their flesh for human food.

The necessity for such supervision was first felt in our export trade in pork products. Our trade with Germany and France had grown to large proportions. The Germans, from their habit of eating pork either raw or only partially cooked, were subject to periodical epidemics of trichiniasis, because the hogs of all countries are more or less affected with the parasite which causes this disease in people. To guard against such outbreaks of disease the German Government instituted a microscopic inspection of all hogs slaughtered in the Empire, so that those infested with trichinæ might be discovered and condemned. Requiring such an inspection for their own hogs, they consider it admissible to prohibit the importation of pork products from other countries where the same precautions are not enforced.

In adopting this regulation it is true that they did not give sufficient weight to the fact that no trichiniasis had ever been produced in Germany by American pork or to the additional fact that the curing process to which all exported meats are subjected destroys this parasite when present. Notwithstanding these facts, the regulation was enforced and our trade was ruined. The agitation which occurred at this time was such that the Government of France was induced to enforce a similar prohibition, although with less justification, because there is no microscopic inspection in that country to exclude the products of their own trichinous hogs from consumption.

Two of the greatest markets of the world being thus shut against our hog products, because of the alleged existence of trichinæ, it is probable that this prohibition will not be withdrawn until our Government provides for an inspection from which it can guaranty that the meats packed under its supervision are free from this parasite. And this is the first reason for the development of a system of meat inspection.

The change in the method of slaughtering by which this business is concentrated in a few centers, and the meat is shipped in cold storage to all parts of the country for sale, has developed the second great reason for national inspection. Under the old plan of killing in the vicinity where the meat was to be consumed the butcher was known by the consumer; he had a reputation to maintain, and he was subject to local laws and sanitary regulations. At present the identity of the butcher is lost in a distant packing center, where several firms carry on an enormous slaughtering business; local butchers who dare to compete are ruined by the power of these great monopolies; the consumer has no recourse but to buy such meat as is furnished or to do without this necessary article of food. Local inspection laws are impotent to protect, because the animal must be seen before slaughter and the viscera must be examined when the carcass is dressed to make an inspection of any real value. This great change in the method of slaughtering animals has made dressed meats an article of interstate commerce, and as such it has been withdrawn to a great extent from local regulations and should receive the supervision of the national authorities.

The third great reason for a national inspection of slaughtering establishments is that of outbreaks of communicable diseases among live stock, which are dangerous to the property or health of our people should be promptly discovered and controlled. The most available and certain method by which this can be accomplished is the inspection of all animals sent to the great slaughtering establishments of the country. Congress has already made large appropriations for the eradication of the contagious lung plague of cattle, and the known infected districts are under strict quarantine. But if the contagion in some way escapes the vigilance of the officers and the disease in spite of all provision for its discovery and suppression appears elsewhere in the country, the owner of affected animals may conceal his losses and ship his herd for slaughter. The animals are purchased by a large packing establishment; they are seen only by their irresponsible employés, and no matter what lesions they show the Government is none the wiser. In this way consumers may be supplied with diseased meats, and the presence of a contagious disease may be successfully concealed. A thorough system of inspection would not only protect the consumer, but would reveal the existence of a new center of contagion at the earliest moment and enable the Government to economically and effectually guard the food supply of the nation and its interstate and foreign commerce from the destructive influence of animal plagues.

To protect the health and lives of its citizens is, or ought to be, one of the first objects of any government, and there is no way in which more can be done in this direction by our National Government than in protecting the food supply of the country. Heretofore this has been left to State and municipal regulations, and the power of the Federal Government has been felt in forcing articles into the several States, because such articles were the subjects of interstate commerce, rather than in supplementing the local authority for the protection of the consumer. And while it is right that interstate commerce should be protected, it is proper that it should be regulated for the benefit of the great mass of the people—the consumers—as well as to maintain the business of the smaller number—the shippers.

The great importance of an inspection of animals slaughtered for food is apparent to the most superficial investigator. In many cities it is a well-known fact that animals in a most disgusting and dangerous condition of disease may be and are regularly slaughtered for human food, and their carcasses go into consumption without let or hindrance. In a country where refinement and civilization have reached the development which we find in the United States the common instincts of humanity require that the consumers of meat should be protected from food of this character.

It is well known that many of the internal parasites of man can only be obtained from the animal food of which he partakes. This applies particularly to tape worms and trichinæ, which pass one period of their lives in the flesh of cattle and hogs. It is not so generally understood that many sudden and dangerous illnesses result from eating animal food which has been poisoned by the products of disease existing in the animal before its slaughter. Most common of these are the septic disorders so common in nearly all species of domesticated animals.

The most dangerous disease in its effects upon human health, however, is tuberculosis, which is known to be very prevalent among cattle, and for which it is seldom that a carcass is condemned. The vital statistics of the country show that 150,000 deaths are caused every year in the United States by this disease, and almost the only available means of prevention suggested by the medical commissions which have investigated the subject, is the protection of the consumer from food derived from tuberculous animals.

In the preparation of this article details have been purposely omitted and the aim has been to present in a succinct form the great reasons for a national inspection law that will guaranty the character of the animal food sold in this country, as well as of that which goes into our export trade. The facts which are here stated, in a broad and general way, are known to the writer to be true, and they can be substantiated at any time by a proper inquiry.

The disease referred to in the concluding part of the preceding paper, tuberculosis, is so widespread and furnishes in itself so strong an argument in favor of a rigid national inspection law, that I think it well to introduce it more prominently in the body of this report, and to that end I submit the following extracts from the able address of Dr. G. C. Faville, professor of veterinary surgery, in charge of the United States Bureau of Animal Industry, Baltimore, Md., delivered at Grange Camp, Va., August 20, 1889. Dr. Faville said:

The close competition in the commercial world and the increasing number of consumers in our cities increases the temptation to adulterate food products; and as the producers of all the foods consumed are really the farmers the question of food adulterations comes home to them through their pockets with great force.

When the farmer's hog has to compete with the cotton-seed oil and paraffine it is expected that the farmer should "squeal," and when his cow has to compete with both the hog and the steer in the production of butter, it is legitimate that he should "kick." It is probable that the chemical processes through which the several ingredients pass in their transformation into so-called lard or butter would destroy any diseased germs that might be present, but when, as often happens, the meat or milk of diseased animals is consumed, the dangers to human life become much greater.

There can be no question that the people of the country do not understand the danger in which they stand, or they would demand protection; and for the people to demand, is for them to secure.

The diseases that are directly communicable to man from the lower animals are numerous: Glanders, anthrax, actynomycosis (lump jaw,) tuberculosis, and others, besides the various digestive and other troubles that come from eating meat from ill-conditioned carcasses, are well known to the medical profession. It is a practice in all the markets of our large cities to sell meat from any animal of the bovine species that is able to go to the slaughter-houses. In most of our markets there is no restriction on the sale of any class of meats, and as a result any old, emaciated cows, or animals with cancerous jaws, bad udders, or any other disease, are slaughtered and sold for human food. While there is much danger from any animal that is not in a healthy condition, the most dangerous of all, probably, is the one suffering from tuberculosis.

Tuberculosis has been recognized under different names for centuries; but not until the last few years has anything like a correct understanding of the disease existed. With this, as with a host of other diseases, the popular name has indicated some existing condition. "Grape," "angleberries," "pining," "consumption," etc., have all alluded to a condition observed in some of the many forms of this disease. It is more than likely that when Moses formulated the Jewish laws against the consumption of animals whose lungs were diseased he referred to tuberculous, and ever since that time the human race has been menaced by this insidious foe.

* * * * * * *

It was stated on the floor of the Senate last winter that statistics showed that five hundred thousand children die annually in our cities in the United States from the use of diseased milk.

What adulterations can possibly be the source of so great danger as the germ of an infectious disease?

While we do not like the idea of buying water for milk, it is surely no more repugnant than buying sand for sugar, or any other adulterated food.

The actual danger to the health of the people from such adulterations is nothing when compared to the danger from tubercular germs.

*　　　*　　　*　　　*　　　*　　　*　　　*

The only way to control the spread of this and other diseases is by compulsory inspection of all cows kept for dairy purpose by competent inspectors, and the condemnation of all cows that are found to be diseased.

A thorough compulsory inspection of all the dairies and slaughtering establishments, under the control of the Federal and State Governments working in harmony, could be made to very materially decrease the number of deaths among the people, more especially in our cities. There is no subject of more importance with which our statesmen can grapple. Under the law as it now stands, the Bureau of Animal Industry can do no more than to incidentally find out that tuberculosis exists; there are no funds that are available for the purpose of controlling diseased meats, or of disposing of tuberculosis cows.

It is not too much to say that scientific men who have given this subject attention are almost a unit as to the necessity of a rigid government inspection of both slaughter-houses and dairies.

The extracts which I append are but a few out of a very large number of a similar character that have been brought to my attention. I will first offer some extracts from the report of an authority already quoted. I refer to the Brooklyn health department. It may be urged in this connection that diseased meats and dairy products are not, properly speaking, adulterations, but it is quite evident that in providing for this report the Committee on Agriculture of the House of Representatives intended to include, as far as possible, all food products injurious to health, as well as those which might come under the head of commercial adulteration. That the committee was desirous to prevent the consumption of diseased meats is shown by their favorable report on the Laird bill which embraced among its provisions one for a rigid inspection of animals intended for human food, both before and after slaughter.

So far as inspection of milk is concerned I would suggest that although from its nature it is unlikely to become the subject of interstate commerce and thus be withdrawn from the absolute control of municipal or State regulations, still there are cases in various parts of the country in which it is conveyed from one State to the other, and it seems to me that to confine the inspection of milk to an analysis as is usually done is quite inadequate for the protection of the consumer. Important as this is, the inspection of the source whence the milk is derived is perhaps even more so. Among milch cows the disease of tuberculosis is undoubtedly very prevalent. Professor Law states that 20, 30, or even 50 per cent. of certain herds that supply New York City with milk are affected by this disease; "in some country districts of New York can be shown large herds with 90 per cent. subject to tuberculosis." The evidence in regard to the contagious nature of tuberculosis and to the possibility of its communication to man by the meat and milk of tuberculous animals is becoming so strong that further doubt is impossible.

Dr. Charles Creighton, of Cambridge University, believes that bovine and human tuberculosis are identical and mutually communicable. He says (see London Lancet, June 19, 1880), 'that he believes a supposed outbreak of typhoid fever which occurred in the industrial school to have been in reality not typhoid, but ' bovine tuberculosis.'"

Dr. Lydten, of Carlsruhe, in his report upon tuberculosis to the National Veterinary Congress, held in Brussels, September, 1883, concludes, among other things,

That there are clinical observations proving the transmission of tuberculosis from animals to man through the use of the milk of phthisical animals; that tuberculosis is contagious, like glanders or lung plague; and that contagion fills a more important role than heredity in the propagation of disease.

Professor Jahne, of Dresden, in a statement of results of all experiments he could collect in 1882 upon the feeding of tuberculous matter to various kinds of domestic animals, states that out of 322 animals thus fed 43.5 per cent. took the disease, 51.1 did not, and in 5 per cent. the result was doubtful.

A large number of experimenters have investigated the transmission of tubercle, and the results are convincing and conclusive. Dr. Kammerer, city physician of Vienna, regards infection of tuberculosis by tuberculous meat and milk as quite as fruitful a source of disease as heridity, to which it is usually traced.

Professor Jahne considers it well proven that the milk of tuberculous cows is as capable of communicating the disease to man as to their own offspring. Stang mentions a case of accidental infection of the son of healthy parents by habitually drinking warm milk from a tuberculous cow. Professor Dunne, of Berne, mentions a similar case that came under his own observation.

Dr. Bartley says that without going further into this examination, we can see the extreme importance of careful inspection of all cows whose milk is to be used as food, and it is unquestionably quite as important that all such animals should be inspected before and after slaughter for human food.

Any disease [says Dr. Bartley], whether contagions or not, which produces a continued febrile condition for a number of days, must be regarded as sufficient cause for the exclusion of the milk of cows so affected.

Dr. J. H. Raymond, health commissioner for the city of Brooklyn, in his report for 1884, says (pages 17 and 18):

We know that when the body is properly nourished individuals are more able to withstand the attacks of disease, and even to escape them, than when from any cause they are in a debilitated condition.

My experience leads me to the conclusion that a very much greater proportion of diseased animals is slaughtered and the meat of the same put upon the market than is commonly believed. During the past summer inspectors have been stationed at the slaughter-houses with reference to this detection of impure meat, and they have thus been enabled to discover and condemn meat which would otherwise have found its way into the market. In one of our large cities, at a not very distant date, scores of immature veal were exposed for sale in public market, and it was the opinion of men competent to decide that some of these calves had come into the world only a day or two before and others had been born dead. The effect of such meat upon the health of those unfortunate enough to eat it can readily be surmised; and here let me say that in my judgment no inspection of meat can be of much value unless it occurs at the slaughter-house and before the vicera has been removed. Many cases of tuberculosis and of contagious pleuro-pneumonia have been detected by the inspectors, who were able to interrogate the lungs before they were removed from the animal, which would have passed a most rigid inspection had an examination of the carcass alone been relied upon.

Messrs. May & McElroy, meat inspectors for Brooklyn, in their report, say:

We have endeavored to check the sale of meats *unfit* for *human* food to such an extent, that we can freely say and without any hesitation that there is less bad meat sold in Brooklyn than any other city of the Union, and for this we are in a large measure indebted to the assistance we have received from our police magistrates in severely punishing offenders, also to the press for publishing the proceedings concerning the same.

INSPECTION AT TIME OF SLAUGHTER NECESSARY.

No inspection of carcasses subsequent to time of slaughter is adequate to insure the safety of the public. Among the various diseases to which animals are liable and which render the meat unfit for human consumption, there are some only to be detected by examination of the internal organs, absolutely necessitating, therefore, an inspection at the time of slaughter of each animal by a competent inspector. This is well appreciated in Europe, and especially in Berlin, where there exist "municipal slaughter and investigation houses," established in 1883.

The following is a record of diseased meat discovered and rejected in these houses; 64 cattle, 2,932 organs, 2 calves, 297 hogs, 3,410 organs, all affected with tuberculosis, 247 hogs with erysipelas, 39 other animals with jaundice, 30 with dropsy, 60 meat in unhealthy condition, nature of disease not stated; in 1,407 cases the microscope detected the presence of tape-worm; * 9,000 unborn calves were seized, which would doubtless have found their way into the market as veal, had the animals been slaughtered without the careful supervision secured by the centralizing of the slaughter business. Finally no less than 199 animals were infested with trichina, and destroyed.

Dr. Strawbridge, at a recent meeting of the Philadelphia County Medical Society, in offering a resolution urging State legislation in form of inspection, said:

Anybody can dump any kind of food in Philadelphia and we must take it, but if we refuse to eat it, we are told that we are not good citizens. Meat ought to be inspected when alive and also during the process of slaughtering. Unless you can inspect the animal alive and also when the internal parts can be viewed, the inspec-

---

* The African negroes who are rich enough to feed on beef, which they eat raw, are largely affected with tape-worm. This is doubtless transmitted from the animal direct into the human system. It may not be amiss to observe that the practice or eating underdone meats is one calling for careful attention from the scientist,

tion is useless. In the inspection of milk the principal thing is to see the cows that give it, so as to see that they are not diseased, and to inspect it at its place of delivery. (*American Analyst*, March 28, 1889.)

From an address read before the State Sanitary Convention of Ohio, by Dr. D. H. Beckwith, of Cleveland, Ohio, member of the State board of health, I take the following:

In Twinsburg, a physician informed me a few months ago that in visiting a dairy farm he found a cow covered with sores and undoubtedly suffering from tuberculosis, where milk was being shipped to market with that of other cows for consumption. Similar cases are noticed in various parts of the country every day.

Thomas J. Edge, a special Government agent in Harrisburg, Pa., says in his report July 21, 1886:

"I am not an alarmist, but if citizens could see the cases of tuberculosis which have come to my notice, they would not allow another session of the legislature to pass without at least an attempt to restrict its spread.

"We have found whole herds and dairies affected to a greater or less degree with this disease. Milk is used from animals scarcely able to stand. That milk from diseased animals reaches your city market is so evident that it needs no demonstration; that no amount of inspection will detect the presence of this disease, is apparent."

This report is not essentially different from those of other agents in other localities. It is only a wonder, in view of all the facts which have been gathered by observation and examination, that any children in large cities ever reach man's estate.

A few days ago I attended a family sick with diphtheria. There was no apparent cause for the appearance of the malady about the premises. Upon inquiring I found that children in other neighboring families were suffering from the same dread disease, and I determined to make investigation. It very soon transpired that all of the affected families were being supplied with milk from the same person, *who sold to no other families in the neighborhood.*

I then drove out to this man's dairy farm and found my worst suspicions more than well grounded. The condition of affairs was simply indescribable. The entire water supply came from a stagnant pond, covered with slime and reeking with filth. Several dead animals were floating in the water, partly decomposed. The stalls where the cows were fed, milked, and housed, were filthy in the extreme. No attempt at ventilation had been made, and the food was of the poorest quality. No worse condition of affairs could have been possible at the farm. A kind Providence only spared the people whom this man supplied with milk.

The dangers to which we are exposed from the use of diseased or putrified meats are very great. The flesh of diseased animals is liable to become putrified sooner than that of healthy animals; and the recent discovery of parasites in meat, which is but partially decomposed, where, in short, putrifactive fermentation has but recently set in, points to their presence as the cause of the severe effect of sickness produced from the reception of meat in this particular stage into the stomach.

Dr. Ashmun, the health officer of Cleveland, informed me that during the past year that from fourteen to fifteen cases were poisoned from eating pork that was infected with trichina.

* * * * * * *

To diagnose trichinosis is no easy task. The symptoms in connection with the discovery of the parasite itself, in the suspected food, and removing a portion of the muscle of the patient and placing it under a microscope, will reveal the real trouble.

The Chicago Academy of Science, a few years since, examined portions of muscles taken from 1,394 hogs in different packing houses and butcher shops in Chicago. They found trichina in the muscles of twenty-eight hogs. From these examinations and observations they came to the conclusion that in the hogs brought to this city, one in fifty was more or less affected with these parasites now on exhibition before you. From the worst specimens they found in Chicago, a person eating an ordinary meal of pork (the specimen containing 18,000 to the cubic inch), would soon become infected with not less than 1,000,000 of young trichina.

PUBLIC OPINION.

I herewith present extracts from sundry letters and reports of gentlemen who as scientists, officials, and publicists may be deservedly regarded as authorities on the subject we have in hand.

- I will first quote Prof. Stephen P. Sharpless, State assayer of Massachusetts. This gentleman is probably one of the best informed men in the country on the subject of food adulteration, and he has been good enough to favor me with advance sheets of a pamphlet entitled "Adulteration of Food," from which I will offer here several extracts, and whom I will also quote again in a subsequent portion of this report, devoted to an enumeration of adulterated articles and adulterants. The professor says:

The statute-books of all nations abound in laws upon this subject, which are practically dead-letters.

This arises from two causes: the first is ignorant, indiscriminate legislation; a law which condemns equally flour or rape-seed and chromate of lead in mustard is soon looked upon with contempt. The second cause is too conservative or too definite legislation. There are many laws which condemn specifically certain adulterations. The adulterator carefully avoids these substances, and substitutes for them others, perhaps not less dangerous, and continues in his way.

Another objection to this specific legislation is the fact that, when an adulteration . is thoroughly exposed, it becomes practically dead, since, if it is known to the whole of the trade, it ceases to be profitable.

But all prohibitory legislation necessarily follows the act which it is intended to prevent, and therefore specific laws against any particular adulteration have but little force, since they come too late to be of any benefit; by the time the law is upon the statute-book the old adulteration has been forgotten, and a new one has taken its place.

England has legislated more on this subject during the past hundred years than any other country, and a careful examination of her laws will serve fully to illustrate what I have said.

A law upon this subject must be simple, easily understood, and general in its application, and it should not attempt to control all commercial frauds, but only such as are directly detrimental to health.

Dr. Beckwith, chairman of the committee on adulteration of food, drinks and drugs, of the Ohio State board of health, has also kindly favored me with the advance sheets of the first annual report of the board, from which I extract the following:

At the present time nearly all the adulterations are mere dilutions and substitutions in the interest of pecuniary gain, as exemplified in the dilution of milk with water, and the substitution of glucose for cane sirups, so extensively practiced; careful research showing in nearly all cases that the presence of absolutely pernicious ingredients is the result of accident and not design. An exception, perhaps, may be noted in the use of alum in damaged flour, but the effects of this adulteration upon the human system are as yet a matter of speculative controversy.

The wisdom of prohibitory legislation can be seen on our side of the water by the results obtained in Canada. The work of examination there began in 1876, when 51.66 per cent. of the articles examined were found adulterated. In six years thereafter, or in 1882, this percentage had been reduced to 25—a remarkable showing, when we consider that the only mode of punishment for infraction of the law has been the publication of the names of guilty parties.

It may be safely asserted that in every locality where the law does not deter from the act adulterated articles are on sale in all kinds of food-supply stores, even the most reputable.

The same authority, in an address before the State sanitary convention, said :

In the matter of coffees, teas, spices, sirups, sugars, and many other articles in daily use, short crops or sweeping changes in import duties do not trouble the consumer in the least. The beneficent manipulators of these goods take the import, be it much or little, and bring the supply up to the demand in their own warehouses by a judicious use of cheap home products. The thrifty housewife knows the cost of a box of spice or a package of coffee with the same certainty that the manufacturer reckons the profits on his sales, and both are content. If said sugars are worth 7 cents in New York, glucose can be had for 3 cents in Buffalo, and it becomes a simple example in proportion to lay before the consumer a prime article at 6 cents, and leave the refiner such a margin for profit as his fancy may dictate. In this instance fraud alone is perpetrated; but the same refiner, to lighten the color of his sirups, employs a salt of tin, which is known to be deleterious to health and therefore dangerous to life.

And in concluding he says :

When we consider that the welfare, the happiness, and the greatest prosperity of a nation depends upon the health and morals of its people, and that unpalatable and irritating foods are the prime causes of very many diseases that flesh is heir to, the imperativeness of entering the field, lance in hand, against this insatiable foe to good living and good temper, food adulterations, ought to be apparent to every one of us. So much has been charged and so much proven by those who have given their time and best scientific knowledge to investigations into the conditions of our food products, that ignorance can no longer be made the excuse for inactivity. The most humble among us may become the strongest in this righteous fight. Play must be given to the impulses which are part of all nations, and not the creature of any condition or profession.

We are too prone to thrust upon the physicians and health officers duties which should be our own.

In this age of progress we can not go back to old time simplicity, when the mistress of the house was the presiding genius of the kitchen, but we can and should examine closely every article of food that enters our doors, and call the attention of the proper authorities to any case of suspected contamination. Let this be done in all our homes and there would soon grow a strong public sentiment against food adulteration.

I appeal to mothers to protect their offspring from the ruthless hand of this destroyer.

I appeal to the economist to enter the lists againsts this despoiler of our homes and depleter of our fortunes. I appeal to humanity to shake off the fetters of the most cruel tyrant and exacting despot the world has ever seen. I appeal to the commercial men all over the country to unite as a band of brothers and discountenance the adulteration of food and drink.

In reply to the circular sent out by me March 29, 1889, and which will be found in the appendix to this report, I received a letter from Dr.

Abbott, secretary of the Massachusetts State board of health, containing some valuable inclosures, and in which he says, in regard to the inspection laws of his State:

I will reply briefly that the laws of this State relative to food and drug inspection were enacted in 1882. Not much active work other than investigation was done till 1884, since which time the board has carried on a very successful line of work under these laws. About 26,000 samples of food and drugs, in all, have been examined, and 395 prosecutions conducted in 80 cities and towns in this State. These related to milk, butter, honey, sirups, sugar, molasses, vinegar, cream of tartar, olive-oil, maple sirup and sugar, spices of various sorts, confectionery, coffee, and various kinds of drugs. Eighty-seven per cent. of the complaints entered (344 in all) resulted in conviction. Several articles of an actively injurious character have been found and their sale suppressed and offenders driven out of the State in some instances.

From one of the inclosures forwarded by Dr. Abbott, consisting of a page from an early report of the State board, I extract the following interesting testimony as to the results of the enforcement of the law:

There can be no question as to the beneficial results of the law as executed by the officers of the board in improving the quality of the food and drug supply of the State, especially in regard to milk and butter—in the former case as relates to the quality of the supply, and in the latter as relating to the proper branding and marking of spurious goods. The extensive correspondence of the health department with wholesale houses outside of Massachusetts also confirms their appreciation of the value of the work done in this State, and also the necessity of furnishing articles of undoubted purity for this market. This is especially true of all classes of drugs sold at wholesale by parties outside the State.

The actual economic results obtained by the enforcement of the statutes relative to food and drug inspection can not be stated exactly. The law is comprehensive and its provisions cover a great variety of articles. Its restraining influence extends outside of Massachusetts to manufacturers sending goods to this market. Such parties appreciate the value of the work done in this State, and also the necessity of furnishing articles of undoubted purity for this market.

From Dr. Hewitt, secretary of the state board of health and vital statistics, of the State of Minnesota, I received a letter, which, though probably not intended for publication, treats the subject of food adulteration in a common-sense manner that ought to command attention. He is no doubt justified in the assertion that there is a good deal of clap-trap in the cry of adulterations, in spite of which he concludes that "there are dangers real, important, and to be guarded against." His letter reads as follows:

I am glad that the Department has undertaken to obtain a report of a popular character of food adulterations.

There is so much clap-trap in the stir which has resulted in the law for food protection in the West this last winter, that I shall be glad to see the truth let into the people.

Alum in baking powder!! was the cry that passed our recent law. Our greatest dangers are in *milk*, and possibly in meat. The cry against milk has been a "bonanza" to the patent-food men, and they now boldly propose to supplement, not only cow's milk, but mother's milk as well, and their sales are enormous. One of them applied to me the other day for a list of the parents of new-born children, which I get monthly from all over the State, so that he could get in early with the patent food. The cry of adulteration and danger in foods is become a watchword for the very frauds themselves.

I hope you will get to the bottom and show the people exactly " where the scare comes in " and where it does not. Even water (spring and deep well) is under the bane of adulteration with sewage, bacteria, etc., and Hyatt and other filterers propose a better plan than the natural one.

The very excess of the cry will soon make people take the natural reaction, as in the story of the "boy and the bear," and then the bear will have the best chance because he is so rarely around when talked about.

There are dangers real, important, and to be guarded against, but they want definition and to be taken from the hands of " business men and experts " trading on them into common knowledge and sanitary supervision.

The Hon. F. B. Thurber, of New York, in a letter to the National Farm and Fireside, which appears in that journal December 7, 1888, says:

There should be a general bill. Piecemeal legislation is not satisfactory, and au executive bureau with an adequate appropriation to see the law is carried out, is an absolute necessity. Laws do not execute themselves; if they did, a police force and the machinery of our courts would be unnecessary,

While there is probably not as much injurious adulteration as the public generally think, there is enough to make a national law desirable, and there is a very large amount of adulteration, which, while perhaps not very injurious, is a fraud both upon the stomach and the pocket of the consumer.

The general principles to be kept in view in such legislation is, is injurious adulteration prohibited, while non-injurious articles, which are adulterated simply to increase bulk or reduce cost, should be permitted to be sold, provided plain notice is given to the consumer of what he is getting.

A very valuable feature of the Laird bill is the official publicity which it proposes to give to the examination of food by the bureau established for that purpose ; for, with official publicity, the competition between manufacturers and dealers will constantly tend to raise the standard, while without such publicity competition would work the other way.

A dealer who desires to sell pure food products now is met with unscrupulous competitors who claim to sell just as good goods as he does at lower prices. If he analyzes their products and exposes them, they raise the cry of " trade jealousy " and his exposures have but little effect ; but with official publicity, the hands of the dealer in pure goods would be greatly strenghtened, and, as above stated, this constitutes one of the most valuable features of the Laird bill. The penalty of exposure is far greater than any other penalty, for it always loses the trade and profit to the manufacturer or dealer in adulterated goods.

The Hon. Erastus Brooks, in an address before the board of Pennsylvania, says :

Take, for example, the simple article of candy, much of which is reported to be made from grape-sugar, glucose, and terra alba, the latter being sold at 1 cent a pound and the former at 4 cents. Where granulated sugar costs, by the barrel, $9\frac{1}{4}$ cents, the cheaper grades of this article may be depreciated in value over 50 to 70 per cent. (Page 363, Pennsylvania report.)

Again, in the same report he says :

It is a public duty to resist all impurities both in the food we eat, the water we drink, and in the contaminated air we breathe in all dwellings and workshops and in all that is around them ; and let me say in speaking alike for the State and citizen that *Principiis obsta* is the only safe rule of action.

Mr. H. Wharton Amerling is the president of the American Society for the Prevention of Adulteration, and in an address delivered by him

I find the following strong presentment of the case from the point of view of physical health :

By adulteration man is made sick, and by it he is prevented a recovery and most foully murdered. Why, thus it is that paupers are made plenty, criminals many, lunatics numerous, and Americans known as a nation of invalids. Men are losing their mental powers, and with them their property and honor. Adulterations more than intemperance dement, delude, and madden to fraud, violence. and murder. Progeny inherits mental taints, and possesses an anomalous desire to do violence to others' property and person; to burn houses and wreck trains. More room for the insane is the general demand. Already $50,000,000 have been invested in asylums, while the cure of the insane costs annually $15,000,000. From 1870 to 1880 there was a gain of 26 per cent. of the total population, while the insane population increased over 100 per cent.

THE POOR WORST AFFLICTED.

People in good circumstances do not suffer a tithe as much as the poor in the cities.· These poor buy a few cents' worth of provisions only at a time. Yet insanity is general and nervous disorders are growing to make it more general. Chloral, morphine, etc., are demanded by the victims of commercial fraud, and insanity or death is the end. The rapidly increasing use of these deadly drugs is frightful, and will be more so if supineness of legislation allows the insidious corruption of our food to continue.

NECESSITY FOR OFFICERS.

No law will be able to suppress adulteration that does not provide officers to detect and apprehend it. Private individuals have not the authority by law to demand of dealers and producers admission to their establishments or power to take articles to inspect, or the time and money to analyze them. The purpose of government being protection, it is not the duty of private individuals to provide the expense of such protection. A citizen may feel that he has been injured by adulteration, but as the expense of analyzation is considerable, amounting in most cases to from $100 to $300, he is unable to prove his injury, and thus unabated murder continues.

Under the law preventing adulteration the English are living longer and better than we. Dr. Foster stated before the English medical society lately, that a man's natural life-time is one hundred years, and that all could live that long if they met with no accident and lived properly. From Dr. Farr's observation of the march through life of 1,000,000 children, he found that the English were living longer than formerly, as there had been a gain of 2¼ years in the average life of the people. He claims that the largest amount of sickness is due to insufficient and impure food, and that the first essential to life is pure food. Further, that hereditary taints may be abolished by proper living, and the body left at death stronger and better than when it was taken up at birth, because the old blood cells, the fibers and epithelic scales being cast off, may be replaced with others better, and eventually untainted.

In concluding this address, Mr. Amerling says :

There are quite as many adulterations as pure articles; we have found food adulteration to average 41 per cent. and drug adulterations 38 per cent. Let us realize that we can never perpetuate our government as long as the very nourishment of the babe is poisoned at its mother's breast.

Mr. F. N. Barrett, editor of the *American Grocer*, thus tersely and strongly expresses the commercial view of this matter, "The gist of the whole thing lies in the right of every man to get what he pays his money for."

A reputable firm of druggists writing from Boston, say : "That the extent of adulteration in that State is 'very small;' our State laws are

so stringent that it will not pay to take the risk." They also say:
"Any adulterants used in medicines would be injurious," and they conclude with the assertion that an experience of thirty-nine years shows
that druggists who adulterate their goods generally lose their trade.

Mr. Elisha Winter is the secretary of the committee on legislation
of the National Pure Food Movement; in a debate before the Retail
Grocers' Association of New York and vicinity between himself and
Artemus Ward, he spoke as follows:

Adulteration plays into the hands of the avaricious few, by giving them the chance
to take more than their share of trade, on account of the low prices at which they
can sell poor goods. Some of the arguments that have been brought up against the
possible operation of this bill remind me of the terrible apprehension with which
some persons endeavored to scare the community when railroads were first suggested.
It was gravely argued that the trains would frighten the cows and they would not let
down their milk.

Recognizing the limit to the situation, we present this bill as the only possible re-
sort; or, in the language of the call: "The only protection the honest retailer can secure for himself is to ask the National Government to supplement the various States
and municipalities by reaching imported commodities, interstate transportation, and
in territories, under the jurisdiction of the United States authorities, the sale of food
products." If the General Government will give the retailer this protection, he may
then work out his own salvation by putting his individual guaranty upon all his
goods and demanding that the State authorities shall then recognize the integrity of
his purpose and give him support, instead of making the present class discrimination,
to free himself from which he is now making this organized national attempt.

The same gentleman, in a circular dated March, 1887, says:

The evils accruing from the manufacture, importation, and sale of adulterated
food, drugs, and medicines are patent to all who have given the matter even casual
attention. This traffic is on the increase, and the detrimental influences arising
therefrom extend to the health as well as the pockets of the people. For both hy-
gienic and commercial reasons it is agreed that a remedy sufficiently powerful to
check the evil must be invoked.

The constitution and by-laws of the Central Association of Retail
Merchants of New York and vicinity, says under the heading of " objects and aims :"

SEC. 7. Protection against the adulteration of goods, fictitious labels, imperfectly
prepared food products, etc.: Legislative reform has been practically null and void
looking to the accomplishment of much reform in the adulteration of goods, fictitious labels, etc. It has been very easy to pass laws, but very difficult to enforce
them. Our proposed State association will have a duty to perform in arriving, if
possible, at a happy medium of judgment upon the merits of these questions. This
association should also demand that with every package of food products shall be delivered the guaranty and designation of the quality of goods therein contained.

DR. HASSELL ON THE NEED OF LEGISLATION.

In concluding this report I submit the following deductions of that
celebrated and eminent English scientist, Dr. Hassell, who more than
thirty years ago was employed by the London Lancet to investigate
the adulteration of food products. Dr. Hassell's work on food adulterants is a standard on the subject, and as his conclusions (which aided

so greatly in securing the passage of the food adulteration act) are
as applicable to the present condition of affairs in this country as they
were thirty years ago in England, I give them in full. The second
paragraph, relative to taxation, is of course not applicable to the United
States at this time:

Legislation on the subject is required—

First. For the protection of the public health. The evidence given before the
parliamentary committee on adulteration proves that the deadliest poisons are daily
resorted to for purposes of adulteration, to the injury of the health and the destruc-
tion of the lives of thousands. There is scarcely a poisonous pigment known to
these islands which are not thus employed.

Second. For the protection of the revenue. This will be readily acknowledged
when it is known that nearly half the national revenue is derived from taxes on food
and beverages. It has already been shown that not long since adulteration was rife,
and it still exists to a large extent in nearly all articles of consumption, both solid
and fluid, and including even those under the supervision of the excise.

Third. In the interests of the honest merchant and trader. The upright trader is
placed in a most trying and unfair position in consequence of adulteration. He is
exposed to the most ruinous and unscrupulous competition; too often he is under-
sold, and his business thus taken from him. It is therefore to the interest of the
honest trader that effective legislation should take place, and not only is it to his interest but we can state that it is his most anxious desire that adulteration should be
abolished. In advocating the suppression of adulteration we are therefore advocating the rights and interests of all honorable traders.

Fourth. For the sake of the consumer. That the consumer is extensively robbed
through adulteration, sometimes of his health, but always of his money, is unques-
tionable. It is, however, the poor man, the laborer, and the artisan, who is the most
extensively defrauded; for, occupied early and late with his daily labor, often in debt
with those with whom he deals, he has no time or power to help himself in the matter, and if he had the time he still would require the requisite knowledge. The subject of adulteration, therefore, while it concerns all classes, is eminently a poor man's
question; the extent to which he is cheated through adulteration is really enormous.

Fifth. On the ground of public morality. Adulteration involves deception, dishonesty, fraud, and robbery, and since adulteration is so prevalent so equally must
these vices prevail to the serious detriment of public morality and to the injury of
the character of the whole nation for probity in the eyes of the world. We repeat,
then, that some prompt, active, and efficient legislative interference is demanded for
the sake of public morality and the character of this country among the nations of
the world.

I will conclude this portion of this report with two notable extracts,
one a speech delivered in the Forty-eighth Congress by the Hon. J. W.
Green and the other from a report in the Fiftieth Congress by the late
Hon. James Laird of Nebraska, respectively.

Said the Hon. Mr. Green:

Who will say that he who stamps and passes off little bits of baser metal than the
standard bullion to put in your pockets is guilty of greater wrong than he who pre-
pares and sells to you base and counterfeit compounds, not to say deadly, to put into
your stomach? Possibly the reason for imposing penalties in the one case and neg-
lecting to do so in the other, is that our ancestors could not realize that human cu-
pidity could prompt such depravity as trifling with the health, well-being, and very
existence of myriads of their fellow-men.

Probably every gentleman on this floor knows what steatite or soapstone is; if not,
I will state that it is a soft, calcareous, easily cut rock, but probably surpassing any

other in weight and density. Presumptively, therefore, not the most digestible article of diet known.

Now, sir, what would be your inference, if told by the proprietor of one of these saponaccous quarries, as I have been, that he finds a ready sale for all the "soapstone flour" that he can grind? And who are your customers? Chiefly commercial millers and sugar refiners.

Mine, sir, was that the information tallied with what I had previously seen in print, that the vile stuff enters largely into our tea, coffee, toddy, sweetmeats, and daily bread. Sir, it behooves those who hear to ponder well. Steatite may be an excellent lining for stoves. I doubt its coequal fitness for stomachs. Hot "biscuit for breakfast," "light bread for supper," was wont to gladden my heart in younger days, for in the house of an honored uncle who raised me, "corn bread" as a rule was the staple staff of life.

Think you that biscuit for breakfast or light bread for supper (Heaven save the mark, how could it have been made light?) would have been as palatable as ashcake or johnny if one of the descendants of Job's comforters had kindly volunteered the information that they were to be made out of nice, white soapstone flour instead of the glorious golden grain grown on the broad acres around me?

It is safe to assume, Mr. Speaker, that were the question put to the leading medical men of the country a large majority of them would decide that the alarming increase of late years in nervous, cerebral, and kidney diseases is directly traceable to the cause assigned, namely, adulterated drinks of all kinds, including vinous, malt, and distilled. Is not insanity fearfully on the increase, as evidenced by the overcrowded bedlams of the land and the mania for self-destruction? Then seek for reason why and find it, too, no less in poisoned beverage than in the growing passion for wild speculation.

But I were derelict to my subject, my constituents, and myself did I close without some allusion to like vicious practice in the make-up of medicine; for, sir, human depravity, with utter disregard of human life, has even dared invade the sacred precincts of the pharmacopœia, to lift the tops of the mystic jars on shelves arranged, and to infuse base substance in their portentous contents, where oft the difference of a feather's weight may involve the mortal life of immortal men. Medical skill is impotent to act and powerless to grapple with fell disease in critical juncture, because by base admixture with medicinals it is at loss to know what measure to prescribe to compass the end desired.

I broadly, boldly make the charge and challenge the refutal of investigation. A distinguished physician told me some years since, in a neighboring city, that probably more deaths resulted directly and indirectly from that source than would from disease if left to itself. Almost every leading government in Europe has stringent laws against adulteration. Of these England has perhaps the most perfect and complete system, and yet it is only of yesterday's growth. Less than thirty years ago Dr. John Postgate, a country physician, seeing the abuses perpetrated by adulterators of every class, took the matter in hand and after years of persistent effort, beginning with only one supporter in Parliament, Mr. Scholefield, and with all the large manufacturers and dealers in Great Britain hounding and denouncing him, succeeded at last in having his ideas adopted as embodied in the adulteration acts of the last decade.

The Hon. Mr. Laird, in his able report presenting bill No. 11266, to the House of Representatives, said:

The work is assigned to the Department of Agriculture for the reason that it is germane to certain work already in progress there.

Then, after referring to the Bureau of Animal Industry and the work of the chemical division of the Department, he adds:

A more important reason for this reference to the Department devoted exclusively to the interests of agriculture lies in the fact that the producer of the food supply is most deeply affected by its adulteration; it therefore appears to be the one best calculated to enforce proper rules which, while accomplishing the objects desired, bear without undue weight upon commerce, manufacture, and transportation.

That one of the first considerations of every civilized government is its food supply can not be controverted, nor can it be contested that the purity of the supply is as important as the supply itself. Quantity alone will not meet the demand; quality, within certain limits, is as necessary to health and the prolongation of life as quantity is to its preservation.

The recent exhaustive examination into the alleged adulteration of lard by this committee demonstrated the prevalence of covetous and dishonest practices in the degeneration, counterfeit, and substitution of commodities by which inferior, cheaper, and sometimes injurious articles were made to represent those of standard quality and absolute purity.

This state of facts amounts not only to a premium upon dishonesty but is a threat to national health. Honest manufacturers and dealers are placed at a disadvantage or are forced into a reckless competition with fraud. Legitimate trade is handicapped and demoralized. It tends to make an Ishmaelite of both manufacturer and dealer, and the hand that is raised against competitors in trade falls in the case of the meat industries of the country necessarily upon the 7,000,000 and over of farmers who produce the supply, and fraudulently upon the entire population that consumes it at second hand.

      *   *   *   *   *   *   *

To say nothing of the home interests to be conserved by the legislation herein proposed, of the protection to health, and the defense against imposition attempted by this bill, the importance of our exports alone is sufficient to require the passage of this act.

The value of farm animals as given by the statistical abstract from 1865 to 1883 shows a steady increase both in number and value. Beginning with a value of $300,879,128 in 1865, they reach the vast aggregate of $2,338,215,268 in 1882. In 1884 the value was $2,467,368,924; in 1885, $2,456,425,383; in 1886, $2,365,159,802. In 1887 there is a shortage of the herds, and only a slightly increased valuation. The statistics from 1884 to 1887 shows proportionally a more alarming decrease, the figures being as follows:

Years.	Number of hogs.	Value.
1882	44,122,200	$263,543,195
1883	43,270,086	291,651,211
1884	44,200,893	246,301,139
1885	45,142,657	226,401,683
1886	46,092,043	196,508,861
1887	44,618,836	200,013,291

For 1887 there is an apparent increase in the total value, but in reality only 21.8 cents per hog, and is due entirely to the fact that the hog crop fell off 1,474,207 in numbers from all causes, and the fact stands forth, leaving out the short crop years of 1883 and 1887, that the decrease in these six years in value of the hog product reached $66,973,301, with comparatively no increase of the numbers.

The opponents of pure lard claim that the admixture of foreign compounds has increased the value of hogs and cattle. The facts prove that within the seven years this compounding has been going on the values have steadily decreased, as shown by the foregoing tables.

      *   *   *   *   *   *   *

The interest in and necessity for the legislation proposed in the accompanying bill is evidently felt most by two classes of our people—the producers and the consumers of the products to be affected. The demand for this legislation is wide-spread among the farmers of the whole country. It is confined to no section and is as emphatic as it is universal. Letters, bulletins, and resolutions from the Grange, Patrons of Husbandry, Alliances, and other agricultural organizations demanding legislation have flooded the committee and the House. That the same interest in the subject is apparent on the part of the consumers is manifest by the letters, petitions, and memorials of the various labor organizations, appealing for such action as will give them honest food as against the dishonest compounds that not only rob them of their money, but of their health also.

*　　　*　　　*　　　*　　　*　　　*　　　*

The interest manifested in this matter by the producers and consumers of the country has received and is receiving the unequivocal indorsement of the trade associations of the United States and England, and has the earnest support of all manufacturers and dealers not directly interested in compounding the products that are the subject of interstate and foreign commerce. That such is the attitude of the business interests of the country, from the least individual to the most powerful trade association, is evidenced by their declarations received by the committee from all sections of the country, and particularly from the great business and food centers of the West and South, namely, Chicago, St. Louis, Cincinnati, Kansas City, Omaha, and other important packing points.

The universality of this demand for immediate legislation should surprise no one. The interests involved are the greatest known to America. The number of horned cattle in the United States in 1885, is put at 45,510,630 head; sheep 50,000,000 head, swine 45,000,000 head, representing in the aggregate $2,500,000,000 [i. e., including horses and mules not given]. This vast sum represents the present earning and possible future profits of half the population of the United States.

In the thirteen Southern States, beginning with Virginia and ending with Texas, and including Kentucky, Tennessee, and Arkansas, all the assessed real estate and personal property, as returned in the census of 1880, did not equal the present estimated value of our animal industry; and all the New England States combined, with the single exception of New Hampshire, did not have enough assessed valuation in 1880 to equal the present value of our domestic animals.

The product of our animal industry in 1884, including meat and labor, and dairy products, and wool, and lard, and tallow, and hides, etc., was four times as much as the gross earnings of all the railroad companies in the United States.

The animal industry is not only great in itself, but it is great in the assistance which it renders to other productive industries. Take the greatest crop produced in this country—the corn crop—and 72 per cent. of that is dependent upon our animal industry for a market. Take the great hay crop, and there is no other way to utilize it; and the oat crop, which mostly goes for animal food. The value of these three crops, which are marketed as animal food, of itself reaches a thousand millions of dollars a year.

While this industry, which asks for the protection proposed in this bill, reaches all the levels of life from the millionaire to the day laborers, it embraces more than all other industries in the country combined—the property of the poor.

One head of this vast aggregate of 45,000,000 of horned cattle is the unit of the wealth of the farmer—it is the savings-bank of the day laborer.

From all these sources come the demand for this legislation, and to its force is added every argument that springs from health, economy, and business honor. To the inestimable damage which delay or defeat must do to the health and morals of the country is added the fact which your committee must in candor state is more than a menace, and that unless it be averted by this or like legislation we are face to face with the indisputable fact that one by one the nations of the earth will close the doors of their trade in our faces, thereby subjecting the great industries dependent on foreign consumption and confidence to irreparable injury.

FOOD PRODUCTS ADULTERATED AND ADULTERANTS COMMONLY USED.

Under this head I will endeavor to give a list of the food products in which adulterations have been frequently practiced, as well as a list of the adulterants and cheap substitutes in common use. I will quote freely from the advance sheets of Professor Sharpless's report, as this gentleman has gone very fully into the subject. As he has already been quoted in the section of this report next preceding, it will be unnecessary to introduce him further to the reader. The same may be said in regard to Dr. Beckwith, of the Ohio Board, from whose report we shall make some interesting extracts on the use of glucose especially, and also as to the percentage of various food products subjected to adulteration. I will also include a list of articles found to be specially liable to adulteration by the chemists of the Massachusetts Board of Health. To this long list we will have to add certain sections treating especially of canned goods and lard adulterations and some notes culled here and there from recognized and competent authorities in regard to the extent and character of these adulterations.

With these notes and figures before him, the legislator will be able to appreciate the extent to which misbranding, sophistication, adulteration, and general debasing of food products is indulged in throughout the country, and it is to be hoped that this section of the report alone will be sufficient to impress upon the minds of all who read it the urgent necessity for adequate national legislation on this subject, which is of such profound interest to every individual of our vast population. Legislation capable of preventing these frauds will mean not only the saving of millions of money to the public, but the preservation of the health of our people.

To quote first from Professor Sharpless. The professor classsifies under the three heads of deleterious, fraudulent, and accidental.

Under the first he includes such adulterations as copper in pickles, red lead in cayenne pepper, arsenical colors in candy, water in milk (deleterious because diminishing the food value of the product and "so starving children who are fed on it)." He also includes in this class of deleterious all sophistications of drugs and medicines, "since the physician depends greatly upon the purity of these in regulating the size of the dose, and if of inferior strength they do not produce the desired effect, and thus become negatively injurious."

The fraudulent he defines as non-injurious, but "a fraud upon the pocket," and to this class belong the great majority of adulterations.

To this class belong such articles as package coffee, which is generally a compound which contains no coffee; salad-oil, which is frequently free from olive-oil, consisting mainly of cotton-seed oil; mustard diluted with flour and colored with turmeric; the mixture of inferior grades of goods with higher grades of the same material, the mixture being represented as pure and of full of strength; the mixture of corn-sirup or glucose with cane-sirup, the mixture being sold as pure cane-sirup; the sale of oleomargarine or suet-butter as genuine butter, and the adulteration of spices with ship-bread.

48

Under the head of accidental, he classes adulterations consisting of
substances accidentally present in articles of food, that is, not added
intentionally, but either present because natural to the article, or because they have become incorporated in it during the process of manufacture. In case such accidental impurities are of a nature to be injurious to health the article should be condemned at once, though in
such case it would hardly be just to hold the vender liable to a greater
extent than is involved in the loss of his property. We append the
following list:

Articles liable to be adulterated, as presented by Professor Sharpless.

Articles.	Deleterious adulteration.	Fraudulent adulteration.	Accidental adulteration.
Arrowroot		Other starches, which are substituted in whole or in part for the genuine article.	
Brandy		Water, burnt sugar.	
Bread	Alum, sulphate of copper.	Flours other than wheat, inferior flour, potatoes.	Ashes from oven, grit from mill-stones.
Butter	Copper	Water, other fats, excess of salts, starch.	Curd.
Canned vegetables and meat.	Salts of copper, lead.	Excess of water	Meat damaged in the process of canning.
Cheese	Salts of mercury in the rind.	Oleomargarine.	
Candy and confectionery.	Poisonous colors, artificial essences.	Grape-sugar	Flour.
Coffee		Chiccory, peas, rye, beans, acorns, chefns-nuts, almond or other nut-shells, burnt sugar, low-grade coffees.	
Cocoa and chocolate	Oxide of iron and other coloring matters.	Animal fats, starch, flour, and sugar.	
Cayenne pepper	Red lead	Ground rice, flour, salt, and ship-bread, Indian-meal.	Oxide of iron.
Flour	Alum	Ground rice	Grit and sand.
Ginger		Turmeric, Cayenne pepper, mustard, inferior varieties of ginger.	
Gin	Alum salt, spirits of turpentine.	Water, sugar.	
Honey		Glucose, cane-sugar	Pollen of various plants, insects.
Isinglass		Gelatine	
Lard	Caustic lime, alum.	Starch, stearine, salt*	
Mustard	Chromate of lead, sulphate of lime.	Yellow lakes, flour, turmeric, Cayenne pepper.	
Milk	Water	Burnt sugar, annotto	Sand, dirt.
Meat	Infested with parasites.		Tainted.
Horse-radish		Turnip	
Fruit jellies	Aniline colors, artificial essences.	Gelatine, apple jelly	
Oatmeal			Old and wormy.
Pickles	Salts of copper, alum.		
Preserves	Aniline colors	Apples, pumpkins, molasses	
Pepper		Flour, ship-bread, mustard, linseed-meal.	Sand.
Sago		Potato-starch	
Rum	Cayenne pepper, artificial essences.	Water	Burnt sugar.
Sugar	Salts of tin and lead, gypsum.	Rice-flour	Sand and dirt, insects dead and alive.
Spices		Flour, starches	
Cloves		Arrowroot	
Cinnamon		Spent bark	
Pimento		Ship-bread	
Tea		Foreign leaves, spent tea, plumbago, gum, indigo, Prussian blue, China clay, soap-stone, gypsum.	Ferruginous earth.
Vinegar	Sulphuric, hydrochloric, and pyroligneous acids	Burnt sugar, water	
Wine	Aniline colors, crude brandy.	Water	Sulphate of pottassa.

*It was evidently an oversight to have omitted cotton-seed oil and water.

As the professor says, the above is certainly a formidable list. Fortunately, however, the majority of articles are not adulterated injuriously. Many adulterants have been only recently met with. Of brandy the professor says :

A large portion of that in the market is made from so-called neutral spirits, which are merely alcohol which has been rectified by passing over wood charcoal. These neutral spirits are colored with burnt sugar, or " French color," and flavored with oil of cognac ; a little catechu is then added, so as to imitate the taste of the wood, and, finally, a little simple sirup, so as to take off the rough edge and impart a smooth taste. In this country the spirit used is generally free from any objectionable ingredients. The foreign article, made from potato-whisky, is more objectionable.

CANNED VEGETABLES AND MEATS.

Frequent cases have been reported of late years of sickness arising from the use of canned meats. The cause seems mainly to have been improper methods of canning, or the use of meat that was tainted before being canned. Unfortunately we can do nothing in such cases by an inspection of the meat, for it generally appears to be all right.

In buying meats and vegetables—if care is taken in their selection, all cans being avoided which are not concave in the heads—but little risk is run.

An examination of the outside of the can is the only guide we can have in this class of articles. The heads should be slightly concave. This shows that they were hot when scaled. If the heads are convex, it shows that decomposition has commenced in the can.

Cheese.—Occasionally cases of poisoning result from its use, but this occurs rarely. The rind is frequently washed with arsenical and mercurial washes to protect it from flies and other insects. All cheese is artificially colored ; generally with annatto.

Candy and confectionery are the subject of such poisonous adulterations and are so largely used by women and children, that we will quote what the professor has to say in regard to them in full, adding that Professor Tonry, in an article in the Baltimore *Sun* last year, made some startling statements as to the extent of the adulterations and poisons used in the manufacture of candy.

Professor Sharpless says :

No article of food is so liable to be injuriously adulterated as candy and all kinds of confectionery. Even the perfectly white candy, which is free from injurious coloring matters, is frequently flavored with fusel-oil (essence of banana), oil of bitter almonds, or essence de mirabano (nitro-benzole), Prussic acid in various forms known as almond flavor, and various other essences and extracts which are poisonous in their nature, and which are used in large excess by the makers in order to give a strong flavor to the article. Various coloring matters of a poisonous nature are used in the colored candies frequently to be found in the shops. A long list of such articles may be found in Hassall's Treatise on Foods, or in an article published in the proceedings of the American Pharmaceutical Association for 1878. The vegetable colors can frequently be identified by dyeing pieces of mordanted cloth with them in a bath slightly acidulated with acetic acid. The aniline colors are easily identified by dyeing unmordanted wool in a neutral or slightly acid bath. Mineral colors must be sought for by the usual methods of qualitative analysis. For identification of coloring matters Bolley's Manual may be studied with advantage.

Glucose is probably present to a greater or less extent in most candy, but frequently the candy is almost entirely composed of it. Such candy should be examined for free sulphuric acid and for excess of lime or sulphate of lime, since the glucoses of the market generally contain an excess of these bodies. The glucose itself is harmless; it is only its impurities that are to be feared. "Terra alba," which may be either gypsum or China clay, is frequently found in certain kinds of cheap candy, such as conversation lozenges. They are to be sought for in such candies as have a very white opaque appearance. Flour is sometimes classed as an adulterant. It is very apt to be present, as it is used for various purposes in the manufacture of candy. It is harmless, and less injurious than the real article. Its use, even in excessive quantities, can only be condemned on the ground that it is a fraud, so far as it is used to make weight. The frosting on cakes, being of the same nature as confectionery, is subject to the same adulterations and frauds.

Very frequently papers colored with Paris green, or with the aceto-arsenite of copper, have been used as a covering for rolls of lozenges, and such papers are almost universally used to make the ornamental leaves with which cakes are ornamented. Such a practice can not be too strongly condemned.

In *coffee*, the adulterants are rarely harmful; fraudulent ones, however, are very numerous. The berry is polished and variously manipulated so as to deceive the customer as to quality, and it is moreover sometimes weighted with water after roasting, by subjecting it to a current of steam while still warm. He adds: "It may be safely said that scarcely a brand of the so-called package coffees contain any coffee." "The essence of coffee," he declares, consists mainly of burnt molasses, while the package coffees referred to, are composed principally of peas, chiccory, and rye roasted and ground. There is also a small nut, "chefus" occasionally found in the market which is used for the same purpose. Almond shells are treated with molasses, and when roasted make a fair imitation.

Cocoa and chocolate.—Of this the chief adulterants are fats other than the cocoa nibs; also flour, with the addition of oxide of iron as a coloring agent to counteract excess of flour. The so-called soluble cocoas have their fat extracted by heat.

Flour, though extensively debased in Europe, is generally pure in this country, though sometimes corn-meal and rice are used as adulterants. The principal trouble is, however, in the use of damaged wheat.

Ginger is sold of two grades, pure and colored, the latter being a mixture of about one-half turmeric. It is used to flavor and color gingerbread. Mustard is also added to strengthen other grades of ginger, so that it may sell as genuine African ginger.

Honey is adulterated with glucose.

Lard is fraudulently adulterated with alum and lime water to improve its color and add to its weight; also sometimes with starch, and largely with cotton-seed oil and stearine.

Mustard is colored with chromate of lead, also with turmeric, and weighted with sulphate of lime.

Horse-radish grated fraudulently consists largely of turnips.

Fruit jellies are often, without respect to the name they bear, simply apple jelly, colored and flavored to suit. The coloring matter is often objectionable, and this is true of most of the flavorings.

In *Pickles*, sulphuric acid and copper are the chief things objected to. *Preserved fruits.*—Apple-sauce is often nothing more than pumpkin boiled in cider. The raspberry jam is often sour, while strawberry jam is frequently made from refuse berries.

In *Pepper* we have roast ship-bread, mustard husks, and Indian-meal. In *Sago*, Potato-starch.

Sugar, generally pure, though powdered sugar is sometimes adulterated with flour and sirups; frequently with glucose.

Of *Spices*, the professor holds that to obtain them pure "it is almost necessary to buy them in the unground state."

Tea is adulterated with other leaves; exhausted tea leaves are said to be shipped from China, while sand and dust are frequently found in the low grades. Teas are weighted between 20 and 25 per cent., and black lead, Prussian blue, and soap-stone are all used.

Good *Vinegar* should contain at least 4 per cent. of acetic acid. By the British law, sulphuric acid and sulphates in vinegar must not exceed one-tenth of 1 per cent., and chlorides of more than one-tenth of 1 per cent. should be absent.

In regard to *Wine*, of which so much is the subject of importation from foreign countries, we will quote what the professor has to say more fully. After defining what wine is, and explaining the process of "plastering," he adds:

This plastering of the wine is excused on various grounds. That it is injurious hardly admits of questions ; but at present it has to be submitted to, since it is impossible to obtain sherry or port that has not been so treated.

The next sophistication to which wine is submitted during its manufacture is sweetening with sugar. In bad years the grapes are poor, and yield a thin, acid juice. In order that the fermentation may produce sufficient alcohol to keep the wine, it is customary to add sugar to the must.

Sometimes the flavor is deficient. In order to improve this, the wine merchant keeps a supply of old, high-flavored wines, which he adds to the poor wine.

If the color is deficient in the case of port and other red wines, elderberry-juice is added to produce the desired shade. In the case of sherries, caramel is used. Champagne is a manufactured wine, having for its basis the juice of a black grape growing in the champagne district in France. The bottlers of champagne flavor it, fortify it, and sweeten it according to private formulas, so as to imitate as nearly as possible their well-known brands; the main difference between the vintages of various years being in the good or poor quality of the wine they start upon. Only one thing can be relied upon in regard to champagne, and that is the fact that it is not the *pure*, fermented juice of the grape. It is just as much an artificial production as the various cordials which are found in the market.

So far only what may pass as reasonably pure wines have been spoken of; that is, wines which come from the places where such wines originate, and which are free from adulteration other than what custom and long usage have sanctioned.

Fictitious wines, that have never been within the limits of the wine-growing districts, are also to be found in the market. It is estimated that the champagne district in France does not produce more than one-tenth of the amount of champagne consumed, the remainder being manufactured from cider and other wines.

Cream of Tartar is frequently more or less adulterated, usually with terra alba or gypsum, and to an extent varying from 5 to 75 per cent. On the subject of *Glucose*, Dr. Beckwith says:

In view of the fact that about ten pounds of this product are manufactured in the United States for every man, woman, and child therein, annually, and that Cleveland is not quarantined against the rest of the country, a two hours' fruitless search for a small sample to be used in comparative analysis was certainly discouraging. Druggists, wholesale and retail, had none, but, with singular unanimity, referred the inquirer to the candy manufacturers, who, to a man, knew nothing of the commodity.

Parenthetically, a specimen of taffy of another kind, abstracted from an inviting pile, yielded 79 per cent. of glucose on analysis.

Glucose is probably the leading adulterant upon the market. It is largely used in sirups, low-grade sugars, jellies, and cheap confections.

As artificially prepared it differs materially from cane-sugar, having but about one-third the latter's sweetening power and being devoid of color when in solution. It is frequently contaminated by the lime which is used to neutralize the sulphuric acid employed in the conversion of the starch into sugar.

Free acid is also claimed to be occasionally found. The presence of lime in the ash of sugars or sirups, obtained by burning off the organic matter and carbon, is a good indication of glucose adulteration in the example under estimation.

The cheapness of glucose, together with its close relationship to cane-sugar, enables the refiner to use it extensively as a sophistication at a handsome profit, and without fear of detection when shipped to country store-keepers and city dealers in localities where sanitary laws are not rigorously enforced.

Per se glucose is wholesome and nutritious. Its production encourages corn-growing, increases the sugar supply, and offers employment to capital and labor. So far the industry which it represents should be encouraged, but its sale as cane-sugar is a fraud, and should be as completely tabooed as the sale of artificial butter for the genuine article. Its cheapness and poverty in saccharine matter demand its complete isolation.

In Europe glucose is largely prepared from potatoes; in this country, on account of its greater abundance, from corn; whence the name corn-sugar or corn-sirup in common use.

Seven examples of sirups were purchased for analysis, principally to determine the extent of glucose adulteration.

These were bought at as many different retail stores in various parts of the city, and were in each case guaranteed free from adulteration. With Fehling's solution six of them were found to contain abnormal amounts of glucose.

The seventh, dark and unprepossessing in appearance, contained less than 4 per cent.; about the usual quantity normal to cane-sugars. The following table shows the complete result:

Glucose yielded (warranted cane).

	Per cent.
No. 1	23.19
No. 2	19.76
No. 3	29.30
No. 4	43.06
No. 5	3.97
No. 6	16.15
No. 7	31.68

Two of these, numbers two and six, the lightest colored in the group, were incinerated, and tested for the salts of tin, which are quite often used as bleaching agents, with the following results:

No. 2, 26$\frac{690}{1000}$ grams contained 231 milligrams of ash and a free precipitate of iron sulphide, which was not weighed.

No. 6, 13$\frac{940}{1000}$ grams contained 184 grams ash and 113 milligrams of tin sulphide. The tin was precipitated from the chlorhydric solution by hydric sulphide. (The presence of iron in No. 2 is not accounted for.) The chlorides of tin are poisons, being classed as such by Taylor and others.

There are many well-authenticated cases of poisoning of people through the medium of the milk of cows which have eaten hyssop, spruce, buckeye (Æsculus glabra), and other poisonous substances, and it is universally known that milk instantly reveals any change of feed having a peculiar or easily-recognizable taste or smell, like turnips or wild parsnips. In the light of these facts, what may be said of the intentional impoverishment and contamination of milk by the feeding of distillery waste, brewers' grains, glucose, and garbage, which is openly practiced in the city of Cleveland by unscrupulous venders?

An examination of the reports of analysts employed by the Boards in different States will reveal the fact that about all of our food supply is largely adulterated. Household articles, according to the following table, compiled by Dr. Newton, of New Jersey, suffer to an alarming extent. The table is appended:

	Per cent.
Spices and condiments	66
Ground coffee	45
Tea	48
Lower grade sugar	20
Sirups	50
Milk, when not inspected	50
Flour	none
Bread	2
Cream tartar and baking-powder	44
Butter (substitution of other fats)	40
Vinegar	rarely cider
Olive-oil	60

If figures do not lie and scientific research deceive, we surely have in this table a finger-board to the path of professional duty for this and other health boards throughout the United States.—(Dr. Beckwith, in The Sanitarian for July, 1887.)

The following is a list of articles especially liable to adulteration, and adulterants used, according to the report of the chemists of the Massachusetts State Board (see page 24, Massachusetts Report, 1886):

Milk.—Addition of water and coloring matter and abstraction of cream.

Butter.—Substitution of foreign fats and addition of coloring matter.

Spices.—Addition of starch and other foreign powders. Especially true of mustard and pepper.

Cream of tartar.—Substitution of starch, gypsum, and other cheap substances.

Baking-powders.—Alum and other injurious ingredients. Baking-powders have no legal standard, other than freedom from harmful ingredients.

Lard.—Presence of cheaper fats and oils.

Olive-oil.—Substitution of cheaper oils.

Jellies and preserved fruits.—Substitution of cheaper fruits and addition of coloring matter.

Vinegar.—Absence of the required amount of acetic acid and addition of coloring matter.

Honey.—Substitution of cane-sugar, glucose, and other substances.

Molasses.—Addition of glucose, presence of tin or other foreign substances.

Sugar.—Glucose, poisonous coloring matter.

Maple-sugar and sirup.—Glucose.

Confectionery.—Terra alba, poisonous coloring matter, fusel-oil, arsenical wrappers.

Coffee.—Mixture or substitute of various cheaper substances.

Canned fruits vegetables and meats.—Metallic poisons.

DRUGS. FORM OF ADULTERATION.

Opium and its preparations, especially powdered opium, and tincture of opium. Deficiency in the required strength of the morphia.

Cinchona, quinine and its preparations, especially the citrate of iron and quinine.

Quinine pills.—Deficiency in weight.

Compound spirits of ether.—(Hoffman's anodyne). Absence of its most important ingredient, the ethereal oil, or substitution therefor.

Spirits of nitrous ether (sweet spirits of niter).—Deficiency in ethyle nitrite.

Salts of bismuth.—Essence of arsenic.

Tincture of iodine.—Deficiency of iodine.

Iodide of potassium.—Excess of chloride or other impurities.

Bitartrate of potassium.—Excess of lime or other impurities, and substitution of starch and other ingredients.

Jalap.—Deficiency in required strength.

Cochineal.—Loaded with heavy foreign powders.

Essential oils.—Adulterated with turpentine.

Pharmacopœial wines and liquors.—Excess or deficiency in required strength of alcohol and excess in solid residue, addition of water, alcohol, or sugar.

Out of 288 samples examined by Dr. Davenport 132, or 45.8 per cent., were found not to be of standard quality.

The Brooklyn Board of Health published the result of sundry analyses made in 1887, from which we quote the following :

Samples.	Name.	Result.
17	Lager beer for salicylic acid	Seven contained it.
2	Ice-cream coloring	Aniline green.
1	Ice-cream flavor	Essence bitter almonds.
2	Cans of mackerel	Tin and lead found.
11	French peas	Copper found.
3	......do	Copper determined.
1	Milk color	Annatto.
5	Macaroni	Saffron, turmeric, and Martin's yellow.
3	......do	Same as above.
2	Cranberries	One spoiled.
1	Canned beans	Colored with copper.
4	Candies	Two contained chrome yellow.
2	Gelatine	One putrid.
1	Canned corn beef	Negative.
1	Canned corn	Spoiled.
4	Candy colors	Not given.
2	Coffee colors	Not poisonous.
65	Lactometers tested	Fifteen condemned.

Six other articles were analyzed, but nothing was found wrong about them.

From the report of the chemist of the American Society for Prevention of Adulteration of Food we give the result of examinations made by him and compared with similar work by a number of other chemists :

From such examinations he made the following average of the per cent. of adulteration of the more common articles of food and drugs: Olive-oil, 60 per cent. ; castor-oil, 20 per cent.; blue ointment, 61 per cent.; tincture opium, 58 per cent.; spices and condiments, 66 per cent. ; candies, 33 per cent. ; sirups, 50 per cent. ; cream tartar, 40 per cent. ; baking-powder, 44 per cent. ; butter, 40 per cent. ; bread and pastry, 15 per cent. ; milk, 40 per cent. ; lower-grade sugar, 20 per cent. : lard, 70 per cent. ; tea, 40 per cent. ; ground coffee, 49 per cent. ; cider vinegar, 80 per cent. ; ice cream, 55 per cent. ; chocolate, 38 per cent. ; honey, 24 per cent. ; wine, 40 per cent. , beer, 45 per cent. ; spirituous liquors, 33 per cent.

Boracic and salicylic acids.—The use of these acids as "preservative" of liquors and food products urgently demand attention of our legislators, as they have already secured that in France and Germany, where their use has been absolutely prohibited, except—and I desire to emphasize the exception—*on export goods.*

Dr. Abbott, of the Massachusetts Board of Health, says:

While they (these ingredients) are not named as active poisons by authorities on toxicology, there can be but little doubt that their use in considerable quantities, or for a long period of time, would have injurious effects. The value of food depends very much upon the readiness with which it is assimilated in the process of digestion. This process is mainly a destructive one, and anything which retards such a process outside of the body will also have a similar action within it, and hence necessarily impairs to some extent its nutritive value. To this effect should be added the effect of the drug itself upon the human economy.

Professor Gossman objects to the use of salicylic acid for the preservation of butter, and Prof. L. B. Arnold says:

It is not advisable to use boracic acid or salicylic acid in butter. They are objectionable as being foreign substances. They are of no use in the human economy. They neither produce warmth, nor make fat, flesh, or bone. They are medicinal and turn nature out of her course, and it causes a needless expenditure of vital force to absorb, circulate, and cast them out of the system.

The following articles are frequently used in the adulteration of liquors: Indian cockle, vitriol, grains of Paradise, opium, alum, capsicum, copperas, laurelwater, logwood, bazil-wood, cochineal, sugar of lead.

Dr. E. Vallin, in the bulletin of the French Academy of Medicine (volume 16), says that a committee of that body recommended, owing to the difficulty of deducting the exact amount used and the danger of excessive quantities being used, if allowed at all, "that the addition of salicylic acid or its compounds, even in small amounts, in articles of food or drink, shall be absolutely prohibited by law."

On page 48 of the Brooklyn report for 1887 we find the following:

From the facts here stated, I am of the opinion that it is time that the addition of salicylic acid to articles of food receive a check at the hands of sanitary authorities.

Various examinations of Brooklyn and Western beer revealed the presence of the injurious acid. A supplementary report to that above quoted gives a number of extracts from the American Analyst and others, concluding with the recommendation that the use of the acid be prohibited by law, and the writer adds:

Beers that show signs of decay for some reason or other, and which by the use of salicylic acid could be preserved, are not proper beverages for the public.

Canned goods.—A New York canned-goods firm assures us that there is no adulteration in the canned-goods trade. The Can-maker's Protective Union, on the other hand, claims that such adulterations exist, and that acids are used and machine-made cans with injurious results. A Philadelphia firm, dealing in canned goods and dried fruits, writes:

In canned goods, dried peas and dried Lima beans are soaked and represented as fresh by unprincipled canners and dealers. In fruits there is an excess of water used to make weight and fill up the cans.

The cost of soaked peas and beans is about 45 per cent. less than that of the fresh articles; while, in addition to water, we find that "alcohol and molasses" are sometimes used to sprinkle dried fruits and to increase weight, but not often, water being the cheaper.

Prof. S. B. Sharpless, already so extensively quoted, thus speaks in regard to this subject of canned goods in his work on Food Adulteration:

Sometimes, through careless soldering or the use of terne-plate in making the cans, the articles preserved become contaminated with lead. As this, at the most, only exists in very small quantities, its detection is often a matter of difficulty. The best method of proceeding is to destroy the organic matter either with aqua regia or chlorhydric acid and chlorate.

Copper is also occasionally found in these goods; it comes from the copper vessels used in their preparation. This may be detected by the same means that are used in the detection of lead and tin. Copper is to be particularly looked for in canned vegetables and pickles, which were formerly very generally colored with salts of this metal. Another fraud practiced in these goods is dilution with water or with sirup, the can having comparatively little solid matter in it. There has also been frequent complaint of light-weight and small-sized cans.

Crackers—From replies received from cracker manufacturers in the various sections of the country it would seem as if this trade had absorbed a large percentage of the few honest men in the country. All letters received on the subject from manufacturers disclaim any knowledge of adulterants, and one in particular, after vigorously disclaiming any knowledge of adulteration, concludes by authorzing me to "state positively that I use no adulterations in my goods." Adulterations in these goods consists mostly, if adulteration it can properly be called, in the use of inferior grades of flour and butter.

Flour is, as a rule, free from adulterations, though not infrequently mixed with inferior grades, while corn-meal is sometimes added; this, however, is a deception easily detected. 1 think we may set down the adulteration of flour as rare; nor have I been able to find any known instance where soap-stone was used, as recited in the speech of the Hon. Mr. Green elsewhere quoted.

Fish are a food product which hardly admit of adulteration, though it is alleged by some fish packers that fish caught by gill-nets are often placed on the market in an unfit condition for food. Doubtless this applies to fish caught in other ways as well.

The liability to fraud in packing fish is in the packing of stale fish, while a good deal of commercial fraud is practiced by the putting up of inferior and cheap fish, which when packed are sold under the name of a superior and scarcer kind.

Lard.—Since this report was commenced the Chemical Division have issued part 5 of Bulletin 13, relating entirely to lard. Although much data on this particular subject has been carefully collected, it is deemed unnecessary to go further into the subject in this report. The reader who doubts that this important article of daily consumption is adulterated in the most *reckless and outrageous* manner, is referred to the above Bulletin and the report of the Agricultural Committee of the House of Representatives of the Fiftieth Congress, and also to the exhaustive and able report of Hon. E. H. Conger, of Iowa, in presenting the bill to prevent the adulteration of lard.

Flavoring extracts are used more or less in every family, and the evidence I have been able to gather leads me to the conviction that a very small proportion of the flavoring extracts sold on the market is true to name. Most of them are mixtures of acids and other drugs; indeed, the manufacturer who makes a thoroughly pure article has no chance in the market, where the cheapness of the artificial article gives it full sway. The acids and drugs used in the preparation of these goods are all more or less harmful, and as a result we not infrequently come across wholesale poisoning from their use. See for example the Washington *Star*, September 12, where a hundred persons at a wedding were poisoned, presumably by eating ice-cream. A manufacturer of this class of goods writes as follows:

Fully 75 per cent. are sophisticated, and the cost of manufacture reduced about 50 per cent. That of this the retailer gains 25 per cent., while the consumer receives no benefit whatever, and further, that these goods are always sold as pure.

Tea, like coffee, is used in every household in the country, and has long been recognized as a necessary article of diet. Like coffee, too, it has become the prey of the adulterater, with this difference, that the adulterants in the case of tea are frequently not simply commercial frauds, but are often injurious to health, and sometimes extremely poisonous. ·

I subjoin extracts from replies received from two different tea houses, differing widely in their statements.

Chase & Sanborn, Boston, say:

There is very little adulteration in teas and coffees. Teas certainly are pure, and the advent of the coffee-mill in the grocery stores has done away with ground coffee, and thereby with the opportunity to adulterate.

They conclude their letter with the statement that "tea and coffee, as at present furnished, are to a very large extent pure products." But it seems that tea-dealers, like doctors, differ.

Martin Gillet & Co., of Baltimore, say:

Two-thirds of green teas from China and one-half of all Japan teas are faced or colored to give them a deceptive appearance. The facing is Prussian blue, gypsum, soap-stone, plumbago, and other chemicals. They are certainly not beneficial to health. Strange to say, the adulteration adds to the cost; that is, the color or facing costs. It simply enables the unscrupulous dealer to deceive the ignorant buyer. As a rule, all tea shipped from China and Japan are branded by orders from America with the names of the highest grades. No one brands the truth on the poorest grades.

In a letter accompanying the circular this firm says that as far back as ten years ago they opened correspondence with the Commissioner of the Agricultural Department, urging that something be done to prevent adulteration of tea, as they believe the honest dealer would thus be greatly helped, and that it would be the means of increasing largely the use of the most wholesome beverage the world knows.

Vinegar.—Corn vinegar is sold as cider vinegar. The fraud consists in the white wine or corn vinegar being sold for pure cider. The coloring used is made from caramel or burnt sugar. I learn from various sources, quite reliable and in no way connected with the manufacture of either cider or corn vinegar, that the latter is an excellent article for keeping pickles and for table use, but when sold under a false name it adds one more to the many frauds practiced upon the people.

Water.—While more disease is caused and spread in all probability by the use of impure water than by any other one cause, it does not come within my province to discuss that matter, further than to call attention to the fact that many spurious mineral waters are sold, and it is claimed that contaminated water from city wells is largely used in the manufacture of these so-called beverages, because of their ability to take larger charges of gas. The question of impure water, however, is one of so local a character as to call for legislation within each State for itself. It is clearly outside of the province of Congress to consider it.

Concentrated lye.—It is estimated by a manufacturer of acids, soda-ash, etc., that a large amount of concentrated lye is adulterated with salt to the extent of 35 or 40 per cent., the standard being 60 per cent. caustic soda, while the average is actually less than 40 per cent. While this extract is not directly in line with food products, I refer to its adulteration because it is an article in general use in every household, and as illustrative of the general spirit of adulteration extending into nearly every department of commerce, including the shoddy used in our clothing, the poisons used in dyeing various articles, notably stockings, from the wearing of which sickness and in a few cases death has resulted.

APPENDIX.

ADVANTAGES OF INSPECTION.

As an evidence of the advantages of "inspection laws," it may be cited that in 1883, when the milk inspection laws were first put into operation in Massachusetts, the samples examined showed 77.5 per cent. of adulteration, leaving only 22.5 per cent. above the required standard. In March, 1884, the samples from the same cities and towns and a few smaller towns showed 55.6 per cent. above standard, when in 1885 and for the fourteen months ending May 31, 1886, the result showed 66.7 per cent.

To continue the illustration it is only necessary to show the increase in purity as shown by the report, page 82:

Article.	1884.	1886.
	Per cent.	*Per cent.*
Vinegar	30	62
Cream of tartar	66	76
Black pepper	21	59
White pepper	35	50
Mustard	35	43
Ginger	76	83

PETITIONS AND RESOLUTIONS.

The National Grange Patrons of Husbandry and a large number of State and local Grangers have passed resolutions and forwarded petitions urging Congress to pass a pure-food bill; also alliances, wheels, and clubs in all sections of the country have urged the adoption of this measure. It is useless to attempt to enumerate the various trade organizations that have adopted resolutions favoring legislation, but the petitions referred to heretofore in Mr. Laird's report are very numerous, and signed by thousands of the best people in all parts of the land. Since the adjournment of Congress petitions have been constantly sent me from all parts of the country to be presented to that body when it meets in December.

THANKS.

I desire to return thanks to the following gentlemen for courtesies and information extended in the preparation of this report: Dr. Davenport, of Boston, State analyst; Prof. H. H. Webber, professor of agricultural chemistry Ohio State University, Columbus, Ohio; Dr. Balch, secretary New York Board of Health, for valuable reports, and to all others who have kindly assisted me, but whose names are necessarily omitted for want of space.

CIRCULAR TO THE TRADE.

UNITED STATES DEPARTMENT OF AGRICULTURE,

UNITED STATES DEPARTMENT OF AGRICULTURE,
DIVISION OF CHEMISTRY,
Washington, D. C., —— ——, 1889.

DEAR SIR: Having been appointed a special agent of this Department to prepare "a report of a popular character on the extent of the adulterations of food supplies," I am anxious to learn from all reliable sources possible—

First: The extent of adulterations.

Second: The extent of adulterations injurious to health.

Third: The extent of sophisticated goods that are sold as pure.

I am anxious to present a report based upon facts, not conjecture, which will do no injustice to any person or trade, and at the same time show favor to none, and present a true statement of the condition of the matter treated; therefore I ask the various trades engaged in the manufacture of food products to aid me with such data as they can furnish, so as to enable me to produce such a statement as will be of service to the commercial and agricultural interest of the country.

I inclose an addressed and franked envelope for reply.

Trusting you will aid us as far as you can conveniently, I am

Yours, respectfully,

ALEX. J. WEDDERBURN,
Special Agent.

Please state your name, —— ——, address, —— ——, business, ——.

To what extent, if any, does adulteration exist in your trade? ——.

State what adulterants are used ——.

Are they, or any of them, injurious to health; and, if so, which? ——.

What is the estimated adulteration in your trade, what percentage of reduction does it make in the cost of the articles? ——.

What is the relative reduction of the sophisticated articles to the retailer, and what is the reduction to the consumer? ——.

Are the adulterated articles branded true to name, or are they misbranded and sold as pure goods? ——.

Can you refer me to any positive evidence of adulterations or misbrandings of food products in your own or other trades? ——.

Will you kindly give me any further information of a reliable character that will enable me to properly represent your trade in my report? ——.

LAWS RELATING TO ADULTERATION.

New York passed a general law in 1881.

Michigan passed a general law in 1881.

New Jersey passed a general law in 1881.

Massachusetts passed a general law in 1882, which was amended and improved in 1886.

The full text of the above laws (except the last) can be found in Part 2, Bulletin 13, Department of Agriculture; as can also be found the Canadian laws. The adulteration act, "Food, drug, and fertilizer."

New York and Massachusetts have special laws relating to dairy products and vinegar.

New York, Ohio, and California have special wine laws.

Twenty-three States have laws against the adulteration of dairy products.

Illinois and Maine have stringent laws against the sale of adulterated lard.

The United States have laws regulating and prohibiting the importation of teas and drugs not of a standard quality, also regulating the sale of oleomargarine.

The Fiftieth Congress adopted an anti-adulteration law for the District of Columbia.

The British laws can be found in Hansell.

There are stringent laws in Germany, France, Belgium, Austria, Holland; and most of the European Governments require the proper branding of food and the inspection of animals sold for food both before and after slaughter.

The new law of New York relative to the standard of vinegar went into effect on August 27, 1889.

The laws of the State of Massachusetts require that milk shall contain over 13 per cent. of milk solids; "to contain less, or to contain less than 9.3 per cent. of milk solids, exclusive of fat, shall not be standard, except in May or June, when 12 per cent. shall be standard.

Massachusetts law provides for the publication in two newspapers of the names and residences of persons convicted of selling adulterated milk.

To kill knowingly a calf under four weeks old in Massachusetts for the purpose of sale subjects the offender to imprisonment not to exceed six months or to a fine not to exceed $200, or both. And all such underage meat is subject to be destroyed whether the vender knew it to be so or not.

No action to recover a debt for an adulterated article can be made in a Pennsylvania court.

To adulterate any liquor with poisons or other deleterious drugs subjects the offender to a fine of $1,000 and imprison ment not over a year.

MAINE LARD LAW.

Not having seen the following law in print, I give it in full:

[State of Maine, in the year of our Lord one thousand eight hundred and eighty-nine.]

[An act to prevent fraud in the sale of lard.]

Be it enacted by the senate and house of representatives in legislature assembled, as follows:

SECTION 1. No manufacturer or other person shall sell, deliver, prepare, put up, expose, or offer for sale any lard, or any article intended for use as lard, which contains any ingredient but the pure fat of swine, in any tierce, bucket, pail, or other vessel or wrapper, or under any label bearing the words "pure," "refined," "family," or either of them, alone or in combination with other words, unless every vessel, wrapper, or label in or under which such article is sold or delivered or prepared, put up or exposed for sale, bears on the top or outer side thereof, in letters not less than one-half inch in length and plainly exposed to view, the words "compound lard."

SEC. 2. Any person who violates any provision hereof shall forfeit the sum of fifty dollars to the use of any person suing therefor, in an action of debt.

IN HOUSE OF REPRESENTATIVES, March 1, 1889.

This bill having had three several readings, passed to be enacted.

FRED N. DOW, Speaker.

IN SENATE, March 2, 1889.

This bill having had two several readings, passed to be enacted.

HENRY LORD, President.

MARCH 2, 1889.

Approved.

EDWIN C. BURLEIGH, Governor.

www.ingramcontent.com/pod-product-compliance
Lightning Source LLC
Chambersburg PA
CBHW022009190326
41519CB00010B/1450